T0271848

Deep Learning for Healthcare Decision Making

RIVER PUBLISHERS SERIES IN BIOMEDICAL ENGINEERING

Series Editor:

Dinesh Kant Kumar
RMIT University, Australia

The "River Publishers Series in Biomedical Engineering" is a series of comprehensive academic and professional books which focus on the engineering and mathematics in medicine and biology. The series presents innovative experimental science and technological development in the biomedical field as well as clinical application of new developments.

Books published in the series include research monographs, edited volumes, handbooks and textbooks. The books provide professionals, researchers, educators, and advanced students in the field with an invaluable insight into the latest research and developments.

Topics covered in the series include, but are by no means restricted to the following:

- Biomedical engineering
- Biomedical physics and applied biophysics
- Bio-informatics
- Bio-metrics
- Bio-signals
- Medical Imaging

For a list of other books in this series, visit www.riverpublishers.com

Deep Learning for Healthcare Decision Making

Editors

Vishal Jain
Associate Professor, Computer Science and
Engineering (CSE), Sharda University, India

Jyotir Moy Chatterjee
Assistant Professor, Department of IT,
Lord Buddha Education Foundation, Nepal

Ishaani Priyadarshini
Lecturer, School of Information,
University of California, Berkeley, USA

Fadi Al-Turjman
Professor, Department of Artificial Intelligence
Engineering & Director Research Center for AI and IoT,
Near East University, Turkey

Routledge
Taylor & Francis Group

LONDON AND NEW YORK

Published 2022 by River Publishers

River Publishers

Alsbjergvej 10, 9260 Gistrup, Denmark

www.riverpublishers.com

Distributed exclusively by Routledge

4 Park Square, Milton Park, Abingdon, Oxon OX14 4RN

605 Third Avenue, New York, NY 10017, USA

Deep Learning for Healthcare Decision Making / by Vishal Jain, Jyotir Moy Chatterjee, Ishaani Priyadarshini, Fadi Al-Turjman.

Routledge is an imprint of the Taylor & Francis Group, an informa business

ISBN 978-87-7022-389-8 (print)

ISBN 978-10-0084-652-2 (online)

ISBN 978-10-0337-326-1 (ebook master)

While every effort is made to provide dependable information, the publisher, authors, and editors cannot be held responsible for any errors or omissions.

Contents

Sarwath Unnisa, A. Vijayalakshmi, and Zainab Toyin Jagun

**4 Segmentation of MRI Images of Gliomas using Convolutional
Neural Networks**   **77**

*Anupama M. Nayak, R. Rachana, M. M. Vishakh,
S. C. Prasanna Kumar, Praveen Kumar Gupta,
Sumathra Manokaran, and A. H. Manjunatha Reddy*

5   Automatic Liver Tumor Segmentation from Computed Tomography Images Based on 2D and 3D Deep Neural Networks   97

William Tustumi, Guilherme P. Telles, and Helio Pedrini

6 Advancements in Deep Learning Techniques for Analyzing Electronic Medical Records   **133**

K. Manimala, E. Fantin Irudaya Raj, and S. Jagatheswari

7 Telemedicine-based Development of M-Health Informatics using AI   **159**

Jagjit Singh Dhatterwal, Kuldeep Singh Kaswan, and Naresh Kumar

Preface

Healthcare today is known to suffer from siloed and fragmented data delayed clinical communications and disparate workflow tools due to the lack of interoperability caused by vendor-locked healthcare systems, lack of trust relationships among data holders, and security/privacy concerns regarding data sharing. The present generation and scenario are a time for big leaps and bounds in terms of growth and advancement of the health information industry. This book is an attempt to unveil the hidden potential of the vast health information and technology. Through this book, we attempt to combine numerous compelling views, guidelines, and frameworks on enabling personalized healthcare service options through the successful application of deep learning frameworks. The progress of the healthcare sector shall be incremental as it learns from associations between data over time through the application of suitable artificial intelligence, deep net frameworks, and patterns. The major challenge healthcare is facing is the effective and accurate learning of unstructured clinical data through the application of precise algorithms. Incorrect input data leading to erroneous outputs with false positives shall be intolerable in healthcare as patients' lives are at stake. This book is being formulated with the intent to uncover the stakes and possibilities involved in realizing personalized healthcare services through efficient and effective deep learning algorithms.

The specific focus of this book will be on the application of deep learning in any area of healthcare, including clinical trials, telemedicine, health records management, etc. For the book, we are seeking articles that focus on the intersection of deep learning, healthcare, and computer engineering approaches. The book is divided into five sections: Section 1 is deep learning in healthcare; chapters 1 and 2 fall under this category. Section 2 is deep learning in diagnosis and image analysis; chapters 3–5 fall under this section. Section 3 is deep learning in medical and healthcare records; chapters 6–8 fall under this section. Section 4 is deep learning applications in COVID; chapters 10 and 11 fall under this category. Section 5 is deep learning in healthcare cybersecurity and blockchain; chapters 11–13 fall under this section.

Chapter 1 covered the basics of deep learning that relates to predictive analysis, healthcare statistics, and the objective of relating deep learning in healthcare.

Chapter 2 gives a brief review of deep learning, its various architecture, and convolutional neural networks. The authors have also explained how these deep learning technologies can help the healthcare domain to advance.

Chapter 3 reviewed the papers that utilize artificial intelligence on healthcare applications to provide better and remote solutions. Furthermore, the authors concentrated on which type of artificial intelligence technique one can use on what type of problems a person currently facing in the healthcare field.

Chapter 4 aimed to address the need for automatic segmentation of MRI images of gliomas in cancer patients. The DenseNet architectural variant of convolutional neural networks has been utilized to build a highly accurate 3D segmentation tool.

Chapter 5 investigated the tradeoff between computational resources and segmentation quality. The experiments were performed on the liver tumor segmentation challenge (LiTS) database.

Chapter 6 examined the current state of the art, existing deep learning approaches used, the research gap, and recommendations for improved deep learning deployment in EHR research.

Chapter 7 demonstrated a multi-channel m-Health system with a Bluetooth connection that is based on the general packet radio service (GPRS). The authors created a system that uses a smart telephone on a commercially GPRS network to send a patient's biological signals immediately to a hospital.

Chapter 8 discussed the application of artificial intelligence in pandemic management including prevention, diagnosis, treatment, and critical policy decisions in the COVID-19 pandemic.

Chapter 9 aimed to provide an exhaustive systematic review of research publications related to smart healthcare applications using blockchain technology. With a thorough review and study of the research works selected by a systematic approach, this chapter aimed to answer a set of research questions focusing on blockchain platforms, consensus methods, smart contracts, system evaluation methods, and their drawbacks.

Chapter 10 aimed to provide a detailed description of the relationship between blockchain and machine learning techniques to empower an IoT-based healthcare system. This chapter also discussed the major issues and challenges faced while implementing IoT-based smart healthcare systems by collaborating machine learning and blockchain techniques.

We are thankful to all those who have directly and indirectly contributed to this book. We are thankful to the publishing staff for providing us with the opportunity.

Vishal Jain, India
Jyotir Moy Chatterjee, Nepal
Ishaani Priyadarshini, United States of America
Fadi Al-Turjman, Turkey

Acknowledgment

I would like to acknowledge the most important people in my life, i.e., my grandfather Late Shri Gopal Chatterjee, grandmother Late Smt. Subhankori Chatterjee, my father Shri Aloke Moy Chatterjee, my Late mother Ms. Nomita Chatterjee, and my uncle Shri Moni Moy Chatterjee. This book has been my long-cherished dream that would not have been turned into reality without the support and love of these amazing people. They have continuously encouraged me despite my failure to give them the proper time and attention. I am also grateful to my friends, who have encouraged and blessed this work with their unconditional love and patience.

Jyotir Moy Chatterjee
Department of IT
Lord Buddha Education Foundation
Kathmandu 44600, Nepal

List of Figures

List of Tables

List of Contributors

Abbi, Prashant, *RV College of Engineering, India*

Anand, Mayank, *UAE University, UAE*

Arora, Khushi, *RV College of Engineering, India*

Bansal, Sandhya, *Department of Computer Science and Engineering, Maharishi Markandeshwar Engineering College, Maharishi Markandeshwar Deemed to be University, India*

Dattana, Vishal, *Department of Computer Engineering, Middle East College, Sultanate of Oman*

Dhatterwal, Jagjit Singh, *Department of Computer Applications, PDM University, India*

El Kafhali, Said, *Hassan First University of Settat, Faculty of Sciences and Techniques, Computer, Networks, Modeling, and Mobility Laboratory (IR2M), Morocco*

El Mir, Iman, *Hassan First University of Settat, Institute of Sports Sciences, Computer, Networks, Modeling, and Mobility Laboratory (IR2M), Morocco*

Gnanasankaran, N., *Assistant Professor, Department of Computer Science, Thiagarajar College, India*

Gupta, Praveen Kumar, *Department of Biotechnology, R V College of Engineering, India*

Irudaya Raj, E. Fantin, *Assistant Professor, Dr. Sivanthi Aditanar College of Engineering, India*

Jagatheswari, S., *Assistant Professor (Sr), Vellore Institute of Technology, India*

Jagun, Zainab Toyin, *Doctor of Philosophy (Real Estate), Faculty of Built Environment and Surveying, Universiti Teknologi Malaysia, Malaysia*

Kaswan, Kuldeep Singh, *School of Computing Science and Engineering, Galgotias University, India*

Kaur, Arpneek, *Department of Computer Science and Engineering, Maharishi Markandeshwar Engineering College, Maharishi Markandeshwar Deemed to be University, India*

Kumar, Naresh, *School of Computing Science and Engineering, Galgotias University, India*

Manimala, K., *Professor, Dr. Sivanthi Aditanar College of Engineering, India*

Manokaran, Sumathra, *Department of Biotechnology, R V College of Engineering, India*

Mittal, Pooja, *Assistant Professor, Department of Computer Science and Applications, Maharshi Dayanand University Rohtak, India*

Navita, *Ph.D Scholar, Department of Computer Science and Applications, Maharshi Dayanand University Rohtak, India*

Nayak, Anupama M., *Department of Biotechnology, R V College of Engineering, India*

Pedrini, Helio, *Institute of Computing, University of Campinas, Campinas-SP 13083-852, Brazil*

Prasanna Kumar, S. C., *Department of Electronics and Instrumentation Technology, R V College of Engineering, India*

Rachana, R., *Department of Biotechnology, R V College of Engineering, India*

Rajendran, Sindhu, *RV College of Engineering, India*

Ramavenkateswaran, N., *RV College of Engineering, India*

Reddy, Manjunatha, *Department of Biotechnology, R V College of Engineering, India*

Subashini, B., *Assistant Professor, Department of Computer Science, Thiagarajar College, India*

Sundaravadivazhagan, B., *Professor, Department of Information Technology, University of Technology and Applied Sciences Al-Musanna, Sulthanate of Oman*

Telles, Guilherme P., *Institute of Computing, University of Campinas, Campinas-SP 13083-852, Brazil*

Tustumi, William, *Institute of Computing, University of Campinas, Campinas-SP 13083-852, Brazil*

Unnisa, Sarwath, *Research Scholar, Department of Computer Science, CHRIST (Deemed to be University), India*

Vijayalakshmi, A., *Associate Professor, Department of Computer Science, CHRIST (Deemed to be University), India*

Vishakh, M. M., *Department of Electronics and Instrumentation Technology, R V College of Engineering, India*

List of Abbreviations

ABE	Attribute-based encryption
ABMS	Attribute-based multi-signature
AD	Alzheimer's disease
ADA-M	Automatable discovery and access matrix
ADO	ActiveX Data Objects
AE	Autoencoder
AGI	Artificaial general intelligence
AI	Artificial intelligence
AIoT	Artificial intelligence of things
ANI	Artificaial narrow intelligence
ANN	Artificial neural network
API	Application Programming Interface
ASI	Artificaial super intelligence
ASSD	Average symmetric surface distance
AUC	Area under the curve
BAN	Body area sensor networks
BASN	M-body health's area sensor network
BCT	Blockchain techniques
BSN	Body sensor network
BWSN	Body wireless sensor network
CAGR	Compound annual growth rate
CCD	Continuity of care documents
CDM	Common data model
CDRN	Clinical data research network
CGM	Constraint goal model
CLE	Confocal laser endomicroscopy

CNN	Convolutional neural network
CPT	Current procedure terminology
CT	Computer tomography
CXR	Chest x-ray
DA	Discriminant analysis
DBM	Deep Boltzmann machine
DBN	Deep belief network
DL	Deep learning
DNN	Deep neural network
DPKPS	Deterministic pairwise key-predistribution scheme
DSCSA	Drug Supply Chain Security Act
DST-ASPP	Depthwise spatio-temporal atrous spatial pyramid polling
DSTS	Depthwise spatio-temporal separate
DT	Decision tree
DT	Decision trees
DTL	Distributed ledger technology
DUO	Data use ontology
DWI	Diffusion-weighted imaging
DWT	Discrete wavelet transformation
EA	Entity authentication
ECG	Electrocardiogram
EHR	Electronic health record
EI	Entity identification
EKA	EKG-based key agreement
EKG	Electrocardiogram
EMR	Electronic medical record
ET	Enhanced tumor
FCNN	Fully convolutional neural network
FLAIR	Fluid-attenuated inversion recovery
FLOP	Floating point operation
GAN	Generative adversarial network

GDP	Gross domestic product
GDPR	General data protection regulation
GMM	Gaussian model of mixing
GPRS	General packet radio service
GSM	Global system for mobile communications
HCC	Hepatocellular carcinoma cancer
HCI&A	Healthcare informatics systems and analytics
HCN	Hybrid Convolutional Network
HGG	High grade glioma
HR	Heart rate
HTER	Human-targeted Translation Error Rate
HU	Hounsfield unit
ICD	International classification of diseases
ICT	Information and communication technology
IJS	Improved Jules Sudan
IMSI	International mobile subscriber identity
IoMT	Internet of Medical Things
IoT	Internet of Things
IP	Internet protocol
IPI	Inter-pulse interval/interpole interface
ISDN	Integrated services digital network
K-NN	K-nearest neighbor
LAN	Local area network
LBP	Local binary pattern
LGG	Low grade gliomas
LiTS	Liver tumor segmentation challenge
LOING	Logical observation identifiers names and codes
LR	Logistic regression
LSTM	Long short-term memory
LSTM	Long short-term memory network
LW	Low Weight

MCSM	Medical concept similarity measure
ML	Machine learning
MR	Magnetic resonance
MRI	Magnetic resonance imaging
MRM	Medical relatedness measure
MSSD	Maximum symmetric surface distance
MLP	Multilayer Perceptron
NB	Naïve Bayes
NHIA	National health insurance administration
NIBP	Non-invasive blood pressure
NIfTI	Neuroimaging informatics technology initiative
NLP	Natural language processing
NMR	National medical referral
NN	Neural network
NTN	Non-triple-negative
NuSVM	Nu-support vector machine
OCT	Optical coherence tomography
OCT	Optical coherence tomography
OVFKA	Ordered-physiological-feature-based key agreement
P2P	Peer-to-peer
PACS	Picture archiving and communications system
PCA	Principal component analysis
PDA	Personal digital assistant
PET	Positron emission tomography
PFKA	Physiological-feature-based key agreement
PoW	Proof of work
PPG	Photoplethysmographic
PPV	Positive predictive value
PSKA	Physiological-signal-based key agreement
PSTN	Public switched telephone network
PVS	Physiological safety value

Raas	Relay-chain as a service
RAS	Robot-assisted surgery
RBM	Restricted boltzmann machines
R-CNN	Region-based convolutional neural network
ReLU	Rectified linear unit
ResNet	Residual network
RF	Radio frequency
RFT	Random forest tree
RNN	Recurrent neural network
ROI	Region of interest
ROI	Region of interest
RPM	Remote patient monitoring
RvNN	Recursive neural network
SAE	Stacked autoencoder
SARS	Severe acute respiratory syndrome
SBFT	Simplified byzantine fault tolerance
SR	Systematic review
SSL	Secure sockets layer
SVM	Support vector machine
TN	Triple-negative
t-SNE	t-Stochastic neighbor embedding
UCI	University of California Irvine
VD	Vascular dementia
VOE	Volume overlap error
WBAN	Wireless body area network
WCE	Weighted cross entropy
WHO	World Health Organization

1

Amalgamation of Deep Learning in Healthcare Systems

N. Gnanasankaran[1], B. Subashini[2], and B. Sundaravadivazhagan[3]

[1]Assistant Professor, Department of Computer Science,
Thiagarajar College, India;
Email: Sankarn.iisc@gmail.com
[2]Assistant Professor, Department of Computer Science,
Thiagarajar College, India;
Email: mrityunjayarasu@gmail.com
[3]Professor, Department of Information Technology, University of
Technology and Applied Sciences Al-Musanna, Sulthanate of Oman;
bsundaravadivazhagan@gmail.com

Abstract

Deep learning's massive powers are transforming healthcare. In recent years, AI and machine learning have grown in popularity and acceptability. The situation became much more convoluted when the COVID-19 outbreak broke out. During the crisis, we witnessed a rapid digital renovation and the adoption of disruptive technology across different industries. Healthcare was one of the potential sectors that gained many benefits from deploying disruptive technologies. Artificial intelligence, machine learning, and deep learning have all become the most vital mechanisms of the business. Deep learning had a significant influence in healthcare, allowing the industry to progress patient monitoring and diagnosis. Drug development, medical imaging and diagnostics, personalized treatments, and patient monitoring improved the health record management, health insurance, and fraud detection. These are some of the most ground-breaking solicitations of deep learning in healthcare.

Deep learning and machine learning models can process and analyze various medical and healthcare data, both structured and unstructured. Document classification and maintaining up-to-date health records might become manually difficult. As a result, smart health records can be well-maintained using machine learning with its subsection deep learning. With the advent of telemedicine, wearable, and remote patient monitoring, there is now abundant real-time data on health and deep learning which help in perceptively monitoring the patients and predict risks.

Deep learning can efficiently detect insurance frauds and predict future risks. Health insurance benefactors are also an advantage if they use deep learning because the models can predict future trends and behavior to recommend smart insurance policies to the clients.

Natural language processing (NLP) forces deep learning algorithms for classification and identification. These two technologies can be used in recognizing and categorizing health data and can also be leveraged to develop chatbots and voice bots. In the current scenario of telehealth, chatbots play a crucial role. It makes the interface with patients.

This chapter provides a detail summary about the role of artificial intelligence, machine learning, and deep learning in extemporizing the technological improvements in healthcare. It also gives a brief introduction about the various deep learning models, spots out some of the important radiological applications of deep learning algorithms, and finally explains the various applications of deep learning in healthcare.

1.1 Introduction to Deep Learning

Deep learning is part of a broader family of machine learning methods based on artificial neural networks with representation learning. Artificial neural networks are used merely in deep learning, which is a sector of machine learning. These technologies are primarily outstripped for categorization of images, audios, and text.

Artificial intelligence is a smaller group of machine learning, and machine learning is a portion of deep learning. The following are discrete definitions:

- One strategy is to use algorithms to extract insights from data, which is known as machine learning.

- Additional practice is to use a specialized algorithm known as a deep learning concept.

Consider this association in the form of a structure of concentric circles.

Deep learning is a form of an algorithm that aspects to be extremely good at anticipating events. The axioms of deep learning and neural networks are closely swappable.

The cerebral cortex stimulates neural networks. At its utmost primitive juncture, the perceptron is a mathematical model of a biological neuron. Similar to the brain cortex, there can be numerous coats of interdependent perceptrons. Inserted values and elemental data are communicated through this "network" of veiled layers until they extend the output layer. The prediction layer is made up of one or more nodes depending on whether the model only yields a number or whether it is a multiclass classification challenge.

Our elementary data, or input values, stream through the same "network" of hidden levels till they extend the output layer. Depending on whether the model merely outputs a number or if it is a multiclass classification challenge, the prediction layer may comprise one or more nodes.

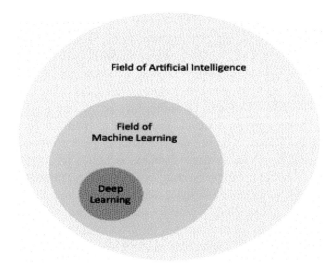

Figure 1.1   The liaison between DL, ML, and AI.

A neural network's hidden units make alterations to the data in order to figure out how it relates to the goal variable. Each node has a weighting that is multiplied by the value of its input. When it does on a number of levels, the Internet is able to transmute the data into somewhat helpful.

The term "deep learning" was devised to have much further concealed units in our neural networks. Similarly, the amount of distinct types of neural networks that can be used has grown. Recurrent neural networks, convolutional neural networks, and protracted precise-term memory are just a few of the popular models.

Deep learning is vital for one reason: it has permitted us to attain substantial, practical precision in real-world situations. Machine learning has also been used to catalogue text and images for decades, but it has succeeded to break past the barrier – algorithms that must satisfy a specific level of accuracy to operate in business scenarios.

A good paradigm of a task that DL has made more feasible for corporate applications is computer vision. DL is not just healthier than any other traditional method for classifying and labeling photos; it is also preliminary to outstrip actual humans.

1.2 Deep Learning in Healthcare

The healthcare industry can now examine data at express speeds while sustaining pinpoint accuracy. Deep learning has a number of benefits in healthcare, including the ability to be fast, efficient, and precise. When it comes to discovering new medicine combinations, deep learning is critical. During the outbreak, disruptive technologies such as artificial intelligence, the subset of deep learning concept, were used to fund the development of vaccines and medicines. Because drug development is a difficult endeavor, deep learning can help make it more efficient, cost-effective, and simple. Deep learning algorithms can anticipate drug characteristics, forecast drug–target interactions, and generate a molecule with the preferred properties. Deep learning algorithms can rapidly process genomic, clinical, and population data, and a multiplicity of toolkits can be used to spot trends. Researchers may now undertake faster molecular modeling and predictive analytics in ascertaining protein structures using deep learning and machine learning [1].

Medical images such as X-rays, MRI scans, and CT scans can be used to train deep learning models. In medical photos, the algorithms may detect the risk and designate irregularities. Deep learning is often employed in cancer detection. Cancer detection via MRI imaging and X-rays has surpassed human standards of accuracy, thanks to picturing recognition. Other common healthcare-related applications comprise drug discovery, clinical trial matching, and genomics [12].

Clinical trials are time-consuming and costly. Predictive analytics using deep learning and machine learning can be used to find potential clinical trial volunteers and allow scientists to pool people from multiple data points and sources. Deep learning will also permit for continuous trial monitoring with minimal human interaction and errors.

It is easier to investigate a patient's health data, medical history, vital symptoms, medical test results, and other information with deep learning models. As a result, healthcare providers are better able to diagnose each patient and give them personalized treatment. Machine learning models can service deep neural networks to prognosis forthcoming health issues or dangers and deliver appropriate medicines or treatments using real-time statistics collected from linked devices [11].

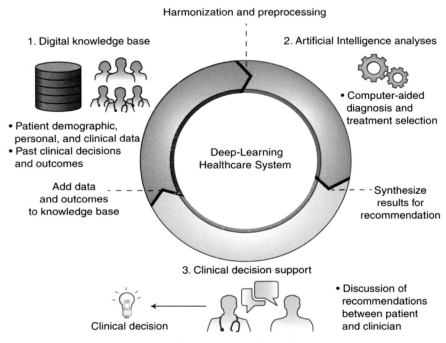

Figure 1.2 Deep learning in healthcare.

Machine learning and deep learning have assisted significant advances in computer vision. It is simpler to treat disorders with an earlier diagnosis through medical imaging.

Over the last few years, artificial intelligence (AI), semantic computing, and machine learning have progressively invaded the medical business. This technical knowledge is linked, bringing something new to the industry and changing the way healthcare experts execute their tasks and provide patient maintenance. They have the proficiency to impact medical technology solutions by adding a new layer to them.

1.3 Artificial Intelligence in the Healthcare System

Artificial intelligence (AI) is aggressively used in the healthcare system with the volume of data nurtures. This includes diagnoses and medication endorsements, victim entanglement, and legislative chores. In many surroundings, AI can perform any type of responsibilities better than humans, but instigating their functionalities may take more time.

Artificial intelligence (AI) and pertinent technologies are becoming very prevalent in business and society, making a successful entry into healthcare. These technologies have the capability to make massive variation and impact in patient care, as well as in many other medical firms [2].

In terms of analyzing harmful cancers and counseling, researchers are able to detect what the inceptions in clinical trials are; algorithms have previously surpassed radiologists.

For example, in the field of radiography, Aidoc has shaped algorithms to accelerate patient diagnosis and treatment. Intracranial hemorrhage, pulmonary embolism, and cervical-spine fracture are among the algorithms that allow the system to rank individuals who necessitate medical attention. This use of AI and deep learning assists the busy radiologist by exposing possible issues, allowing the healthcare practitioner to better regulate and steer patients. By integrating with workflows, it also expands access to vital patient data while reducing paperwork.

1.4 Machine Learning in Healthcare

Machine learning, which is a subset of artificial intelligence, plays a vital role in many health-related dominions, including the development of new medical procedures, the handling of patient data and records, and the treatment of chronic diseases. In healthcare, machine learning is more extensively used, and it is helping patients and medical personnel in numerous ways. In the healthcare industry, machine learning is mostly naturally used to automate medical billing and clinical decision support. In medical industry, machine learning and healthcare approaches are used in a variety of ways. Deep learning in radiology aids radiologists in making refined intuitions when assessing figures such as conventional radiography, CT, MRI, PET scans, and radiology descriptions by automatically detecting difficult patterns.

Automatic detection and diagnosis techniques based on machine learning have initiate that it was perform well as like an expert radiologist. Healthcare machine learning program was experienced to observe breast cancer and attained an exactness of 89%, which is identical to or improved than radiologists. These are just a fraction of the numerous ways machine learning may be used in healthcare [1].

Machine learning in healthcare encompasses forecasting of which treatment techniques are most possible for the patient constructed on the characteristics of the patient. This requires supervised learning, which necessitates the use of a training dataset with a predetermined end variable.

The additional advancement in machine learning is the neural network. It has been used for predicting patients that they will affect from the given disease. The most progressive method of predicting outcomes using deep learning or neural network models with countless levels of features is the modern trend in medicine. There could be a superior number of hidden elements in hottest models, which could be easily recognized by today's graphics processing units

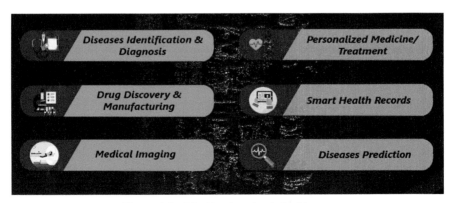

Figure 1.3 Machine learning in healthcare.

and cloud architectures' healthier processing. The most communal application of deep learning is to detect malignant cells in healthcare.

With machine learning induced with deep learning, that can custom neural network models with multiple phases of appearances to predict outcomes is the vital need. Deep learning is frequently used in healthcare to recognize potentially cancerous cells. Machine learning focuses on the large areas such as drug discovery and drug development in pharmaceutical companies.

Recognition of harmful tumors in radiography pictures is one of the common applications of deep learning in healthcare. Deep learning in radiomics is used to catch the clinical patterns in medical imaging which is far from the human's eye. In oncology centered image analysis, deep learning and radiomics can be extensively applied [16].

Some companies such as Quotient Health, Kensi, Ciox Health, etc., use machine learning in healthcare system.

1.5 Natural Language Processing (NLP) in Healthcare

Natural language processing (NLP) refers to a computer's ability to comprehend the most recent human speech words and text. In the prevailing trending technology, it is basically used to provision spam email privacy, personal voice assistants, and language translation apps.

Natural language processing (NLP) systems are recurrently used in healthcare with machine learning. Most NLP-based deep learning in healthcare applications necessitates the practice of medical machine learning. Statistical natural language processing (NLP) is fabricated on machine learning and has facilitated enhanced recognition accuracy recently.

Natural language processing is a profound technique which has the ability to search, analyze, and understand massive amounts of patient data. Using machine learning in healthcare, with the support of NLP, meaningful insights and concepts from data can be extracted which was formerly supposed to be buried in a text form using unconventional medical algorithms. In the latest healthcare innovations, NLP can accurately give a remarkable control to the unstructured data available in the universe, providing great insight into understanding quality, refining techniques, and improving patient outcomes.

The most prevalent applications in healthcare are the creation, understanding, and categorization of medical documentation, as well as published investigation. This system can analyze formless patient healthcare information, generate reports that transcribe patient dialogues, and engage in conversational AI [18].

The most popular NLP techniques that can be used in the healthcare are optical character recognition, sentiment analysis, named entity recognition, text classification, and topic modeling.

1.6 Deep Learning Models

Deep learning is a rising field with applications that span across a number of use cases. It is important to know and understand the different types of models used in deep learning for any of the fields. Deep learning has previously prepared an impact in the healthcare field. Google has placed a lot of work into researching how deep learning models produce predictions about hospitalized patients and outcomes monitoring. The "Deep Learning for Electronic Health Records" blog article exemplifies about how to diminish administrative load when enhancing insights into

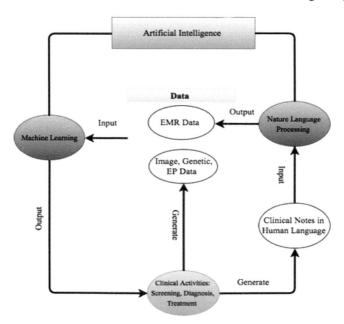

Figure 1.4   Road map from clinical data generation, through NLP data enrichment and ML data analysis.

patient maintenance. Deep learning works finest in healthcare because it drops administrative costs while letting doctors to emphasis on what they fix best: treating patients [15], [16].

Deep learning associations are revolutionizing patient safeguard; it plays a critical part in clinical practice for health schemes. The most popular applications are computer vision, natural language processing, and reinforcement learning.

- Deep learning algorithms can produce exact automatic diagnostic recommendations:
- Lower the cost of care
- Avoid delays in reporting critical and urgent cases
- Free up time for healthcare professionals to focus on more complex diagnoses or patient care by decreasing their administrative workload.
- Auditing prescriptions and diagnostic results to lessen diagnostic mistakes.
- More rapid diagnostics.

With the global outbreak, the use of deep learning models has become additionally imperative. Deep learning applications are presently being investigated by researchers.

Early diagnosis of COVID-19:

- Analyzing chest CT images and chest X-ray (CXR)
- Admission to a critical care unit can be predicted
- Assisting in the identification of potential COVID-19 patients
- Calculating the amount of mechanical ventilation required

The use of deep learning technologies for communication with patients, such as speech and text recognition, will keep escalating in the future. As a result, deep learning applications in healthcare nurture anxieties about data privacy and laws [11].

1.6.1 Interpretation of deep learning models in medical images

Deep learning models' computer power has enabled fast, accurate, and efficient healthcare operations.

The following are the most widely used deep learning algorithms:

- Convolutional neural network (CNN)
- Recurrent neural networks (RNNs)
- Deep Boltzmann machine (DBM)
- Deep neural network (DNN)
- Long short-term memory networks (LSTMs)
- Deep belief network (DBN)
- Stacked autoencoders

By uniting aspects such as tissue size, volume, and form, deep learning algorithms can efficiently assess imaging data. Important segments in images can be emphasized using these models. Deep learning algorithms are utilized to diagnose diabetic retinopathy, Alzheimer's disease early detection, and breast nodule identification via ultrasound.

Deep learning models can assess structured and unstructured data in electronic health records (EHRs), such as clinical notes, laboratory test results, diagnoses, and prescriptions, at lightning speed and with the highest level of accuracy.

Deep learning algorithms can compare prescriptions to patient health information to spot and fix potential diagnostic or prescription errors.

Deep learning models use advanced predictive analytics to assist insurance businesses in making offers to their consumers.

With recent technical advancements, deep learning models have become more important in drug discovery and interaction prediction.

The majority of radiology and pathology photographs will be reviewed in the future. Medical imaging systems such as computed tomography (CT) and magnetic resonance (MR) use image recognition and object detection for picture segmentation, prediction, and illness identification [7].

1.6.1.1 *Convolutional neural networks (CNNs)*
The CNN was the first method for high-dimensional image analysis to be suggested and used. It is made up of convolutional filters that convert 2D to 3D.

Deep learning algorithms make complex data processing easier, letting irregularities to be acknowledged and prioritized with greater precision. Convolutional neural networks (CNNs) provide information that helps doctors diagnose health problems in patients more rapidly and precisely.

CNNs, a kind of deep learning, are especially well-suited for analyzing pictures like MRI data and X-rays. CNNs are intended to process images, which allows them to activate quicker and handle larger images.

Deep neural networks (DNNs), principally convolutional neural networks (CNNs), are frequently used in image sorting tasks that change frequently, and, conspicuously, it has been ascertained to be effective.

With the many CNN-based deep neural networks constructed, the ImageNet Challenger, which is the utmost challenge in major segmentation and image classification task in the image analysis field, was proven to be accomplished. In medical categorization, the CNN-based deep neural system is commonly utilized. Because CNN is such a good feature extractor, using it to identify medical images can save time and money on feature engineering. Furthermore, other research has discovered that CNN-based systems can be accomplished from large chest X-ray (CXR) film datasets and state-of-the-art with high exactness and sensitivity results on their dataset, such as the Stanford Normal Radiology Diagnostic Dataset with over 400,000 CXR and a new CXR database (ChestX-ray8) with 108,948 frontal-view CXR.

CNNs are multilayer computers that use convolution filters to transform their input. CNNs, unlike other deep learning architectures, extract statistics from small parts of input images, known as receptive fields. The input data is convolved with a succession of filters at each layer of this model, each yielding a new feature data.

The pooling layers are used to condense the size of the features by conserving the significant features of the images. The final output of the pooling layer is assumed as an input to the fully connected layers. This implies a set of non-linear functions. Here, the weights are applied for the input vector using the linear transformation method by the neuron.

1.6.1.2 *Recurrent neural networks (RNNs)*

Recurrent ncural network (RNN): A recurrent neural network (RNN) is a neural network design with recurrent connections between hidden states that can study sequences and simulate time dependencies. The recurrent connections are utilized to sense associations across time as well as between inputs.

They are a type of neural network that uses contextual memory to recognize patterns in sequential data. Many sorts of sequential information have been used with recurrent neural networks, including text, audio, movies, music, genetic sequences, and even clinical occurrences.

RNNs, on the other hand, are presently more commonly used in areas of radiology connected to language. Speech recognition software systems, which radiologists employ to write and transcribe reports, are currently their most popular use in many radiology offices. RNNs have also demonstrated the ability to cause text reports to complement abnormality detection methods, as well as heightened the process of disease annotation from electronic health record radiology reports [2], [5].

A deep learning based approach is suggested for decrypting the state of the brain that underpins countless intellectual functions from task of MRI data. The LSTM RNN technique is used to adaptively capture the temporal dependency within the functional data as well as the relationship between the learned functional representations and the brain functional states.

1.6.1.3 *Restricted boltzmann machines (RBMs) and deep belief networks (DBNs)*

This is a unique type of algorithm in neural network which is used in machine learning concept. This is mainly used for dimensionality reduction, regression, etc. This neural network is a two-layered concept and these are associated with using fully bipartite graph.

DBNs are RBM-stacked neural networks in which every layer stores algebraic needs between the units. This concept uses the preceding layer's initiations as contributions in this architecture. The learning method encompasses with two phases that work composed to advance the likelihood of the training data: unsupervised training of the individual layers and final fine-tuning by a linear classifier.

1.6.1.4 *Deep neural network (DNN)*

This involves multiple layers, which lies in between the input and the output layers. We can use different types of neural networks based upon our domain, but it always implements neurons, synapses, weights, biases, and functions.

1.6.1.5 *Generative adversarial network (GAN)*

This design is a class of machine learning framework which is made up of networks called generator and discriminator. The core idea of a GAN is based on the "indirect" training through the discriminator, another neural network that is able to tell how much an input is "realistic," which itself is also being updated dynamically. This basically means that the generator is not trained to minimize the distance to a specific image, but rather to fool the discriminator. This enables the model to learn in an unsupervised manner.

1.7 Radiologic Applications using Deep Learning

The following sections address different radiologic presentations in image classification, image transformation, object identification, image segmentation, image generation, and object detection which helps to learn more in depth about the deep learning applications in healthcare.

1.7.1 Image classification

The task of radiologists is to accurately judge each patient's medical representation, which will analyze the various types of cancer. CNN models are used to reliably scale difficult ailment arrangements since medical pictures encompass a variety of magnitudes and categories of compound disease patterns. A program learning practice that integrates incremental training of more complex topics might confront these tough challenges.

Deep learning entails a massive range of data to generate proper medical pictures. By assembling a large number of deep features from buried layers, the present radiomics model may involve with CNNs approach. The deep features use feature engineering to produce proper analytical pattern for medical images.

1.7.2 Object detection

Object detection is the method used for detecting the boundary of the images. There are two methods used in deep learning for object detection. The area proposal-based algorithm is one of the best cases that could be considered. By means of a selective search technique, this method reveals several types of spots from input photos. Later, the trained model identifies the patches of images and classifies them depending on the region of interest (ROI). The region proposal network was developed to speed up the detecting process.

The regression approach method is a one-stage network that was used to recognize objects in the other methods. These methods were used effectively to improve the pixels of the images and to identify the box coordinates and class probabilities. RetinaNet was recently developed to reduce the complexity in single-stage network. In this network, focus loss was used to overcome the problem of significant foreground–background data or class imbalance. In biological images, per-pixel or coordinated bounding box coordinates with strong labeling are used in various item detections. U-Net and RetinaNet were two important and prominent methods that are implemented to spot the sagittal X-ray of the patients in various parts of the body.

1.7.3 Image segmentation and registration

Since medical pictures include so much data, several deep learning approaches have been inspected and prophesied for application in clinical contexts on image segmentation. It has produced fruitful results in various applications.

In the formal approach of photo segmentation, edge detection screens and numerous scientific algorithms are utilized. In order to improvise the segmentation enactment, techniques like dependent thresholding and close-contour algorithms can be utilized. Another option to examine was the use of registration as a good technique of segmentation. But these methods are very complex and hard to implement. DNNs and CNNs are rapidly being used in medical images to increase segmentation performance. Tumors and other structures in various organs, cells, and membranes have all been attempted to segment separately. In these, patch-based two-dimensional CNN algorithms and post-processing were applied as in normal ML.

To improve medical imaging segmentation performance, deep learning networks have recently been proposed. To generate more precise segmentation results, multitasking with segmentation, classification, regression, and registration works together. Because medical labeling is costly, semi-supervised/unsupervised learning methods based on unlabeled data were suggested. Since these studies have not yet outshined supervised learning's segmentation performance, it is considered a strategy by promise for resolving the large imbalances in medical imaging.

1.7.4 Image generation

It is always a bit challenging to acquire high-quality medical images using CNN in medical industry, and, moreover, capturing medical images and labeling the images take a very long time and are difficult to utilize due to lot of restrictions. To solve the above problems, GAN can be used effectively to develop real synthetic images which will be very useful for CT scans and X-ray.

GAN combines two neural networks to produce accurate artificial images. Numerous medical images such as X-ray, computed tomography, magnetic resonance, positron emission tomography, histopathology images, retinal images, and surgical videography were all created with GANs to perform multiple trials and to produce more stable dataset for exercise classification or segmentation neural networks.

Some of the other important use of GAN is abnormality detection in medical imaging through which the model may try to identify abnormalities gained from the normal photographs. The deep convolutional generative adversarial network model is used in unsupervised inconsistency detection to identify a guiding indicator in OCT images. A study shows that GANs may be used to identify irregularities in medical photographs without the requirement for supervision.

1.7.5 Image transformation

Image transformation study can be very complicated and it may be classified into two important categories: those that used GAN and those that did not.

1.7.5.1 *Without the use of a generative adversarial network, image to image translation is possible*

Style transfer is inextricably linked to the emergence of image-to-image translation. CNN is employed in this process to convey artistic style. Despite their effectiveness in conveying style, these algorithms were rendered impractical due to over-abstraction of characteristics. Wavelet transformation and multilayer stylization can be used to get around this problem. With the addition of a wavelet pooling layer, this research was pushed to the next level, allowing for visually realistic style transmission.

CNN can be actively used to remove image distortion, but also autoencoder (AE) can be used for denoising. Techniques like J-invariant, denoising AE, and layered denoising AE were used to create the Noise-2Self idea. It can be a single image-level denoising approach; so keep that in mind. The CNN network might likewise be used to transmit information on other kinds of transportation [3].

1.7.5.2 *GAN for image-to-image translation*

With GAN, image-to-image conversion with pixel-to-pixel correlation may be performed using conditional GAN. A pix2pix network is the name for this concept. To address the restriction that demands pixel-to-pixel correlation, a cycle GAN architecture was designed that does not entail pixel-to-pixel connection. Although cycle GAN may be used to match mismatched images, style transfer between more than three domains demands a high number of generators – typically, half the number of domains squared. They addressed the challenge by utilizing a single generic generator known as star GAN. Pipeline denoising may be done with GAN. The estimated noise distribution is computed using GAN, and the estimated noise distribution is subtracted from the original picture using another CNN architecture. Making conditional GAN thicker CT to thinner CT is one of the greatest ideas. To make the CT slice thinner, a three-dimensional patch is applied. To produce CT from MRI, the cycle GAN idea is used. However, there is no harmony or data to suggest that utilizing GAN produces better results than not using GAN.

Deep learning, on the other hand, is predicted to aid radiologists in making more precise finding by providing a numerical investigation of suspected grazes and shortening the clinical procedure time.

In recognition and computer vision applications, DL has already demonstrated equivalent recital to humans. Because of these technical advancements, it is plausible to expect some significant changes in clinical practices. Deep learning has already proven human-like performance in recognition and computer vision applications. It is reasonable to predict some big changes in therapeutic practices as a result of these technological improvements. We anticipate that, rather than replacing physicians, AI in medical imaging will act as a collaborative platform for reducing stress and distraction caused by a range of repetitive and tiresome jobs. Deep learning and AI applications in radiology are still in their infancy. There are also a plethora of other issues to address and overcome, including ethical, regulatory, and legal considerations, all of which should be carefully considered when AI is created to utilize clinical image data [7], [19].

1.8 Predictive Analysis using Deep Learning and Machine Learning

Predictive analytics for health data is gaining popularity as a game-changing technology that might lead to more preventive and preventative treatment alternatives. Predictive analytics aids healthcare life sciences and providers by employing a variety of approaches such as statistics, data mining, machine learning, modeling, and artificial intelligence that analyze recent data and generate predictions. It assists healthcare organizations in preparing for healthcare by lowering costs, properly detecting diseases, enhancing patient care, maximizing resources, and improving clinical outcomes.

Predicting future health-related results using clinical or nonclinical patterns is the goal of healthcare predictive analytics. Deep learning applications in drug trials have advanced in recent times and have shown promise in addressing a number of challenges by analyzing the history of the patient in drug discovery and offer the proper medication to the patient based on their symptoms and tests. The predictive analytics in healthcare study outcomes such as hospital readmissions, medical problems, patient death, and therapy responses are frequently of enormous practical value.

Medical imaging procedures including CT scans, MRI scans, and ECG are used to diagnose life-threatening conditions like heart disease, cancer, and brain tumors. Doctors can better analyze ailment and afford the finest treatment to patients. It is also being used to investigate medical coverage fraud claims. It can forecast future fraud claims when used in conjunction with predictive analytics. It aids the insurance business in delivering reductions and offers for target patients. This technology is being utilized to diagnose Alzheimer's disease at an initial stage. Deep learning is a technology for understanding a genome and assisting patients in determining the disease that may impact them, and it has a bright future [1], [7].

1.9 Clinical Trials using Deep Learning

Deep learning technologies are being employed more frequently to improve clinical practice, and the list of instances is rising every day.

Deep learning is used in PET-MRI attenuation correction, radiotherapy, radiomics, and neurosurgical imaging for theragnostic purposes, merging confocal laser endomicroscopy with deep learning models for on-the-fly detection of intraoperative CLE images.

Advanced deformable image registration is another key application field, allowing quantitative analysis across multiple physical imaging modalities and throughout time. Clinical radiologists, including radiomics and imaging genomics (radio genomics), deep learning in neuroimaging and neuroradiology; brain segmentation; stroke imaging; neuropsychiatric disorders; breast cancer; chest imaging; imaging in oncology; medical ultrasound, and so on can all benefit from deep learning [11].

1.10 Applications of Deep Learning in the Healthcare System

Deep learning employs neural networks to analyze a significant amount of data, such as medical reports, patient records, and insurance records.

Deep learning applications in healthcare include solutions for medical imaging, chatbots, and recognizing patterns. Deep learning provides medical professionals with intuitions that allow to recognize problems earlier, allowing them to give much more personalized and significant patient care.

In healthcare, this technology allows clinicians to accurately analyze any ailment and effectively treat patients, resulting in improved medical judgments [11].

1.10.1 Drug discovery

Deep learning in healthcare aids in the discovery and development of new treatments. The technology examines the patient's medical history and recommends the most appropriate treatment. Furthermore, this technology extracts information from patient symptoms and testing [6].

1.10.2 Medical imaging

Heart disease, cancer, and brain tumors are diagnosed by means of medical images using the procedure with a variety of scans and ECG. As a result, deep learning assists doctors in better analyzing diseases and providing the best treatment to patients [7].

1.10.3 Insurance fraud

Deep learning systems can also spot medical insurance fraud by examining fraudulent behavior and health data from a range of sources, such as claims history, hospital-related data, and patient characteristics.

1.10.4 Alzheimer's disease

The major difficulties faced in medical sector are in the Alzheimer's disease. Alzheimer's illness is detected early using deep learning technology.

1.10.5 Genome

In genetics and the insurance sector, deep learning has a bright future. According to Entilic, deep learning techniques are used to make doctors speedier and more precise. A deep learning technology will be very useful for monitoring the health of the children by the parents with the help of smart or digital devices in physical time, by reducing the regular visit with doctors. Deep learning in healthcare can give amazing applications for doctors and patients, allowing doctors to make better medical decisions.

1.10.6 Healthcare data analytics

Deep learning models can analyze structured and unstructured data in electronic health records (EHRs), such as clinical notes, laboratory test results, diagnoses, and prescriptions, at lightning speed and with the highest level of accuracy.

Digital phones and wearable devices afford vital lifestyle facts. They will have the ability to convert the data through monitoring medical risk variables for deep learning models utilizing mobile apps [6] [7].

1.10.7 Mental health chatbots

Happify, Moodkit, Woebot, and Wysa are among the AI-based mental health apps (including chatbots) that are becoming more popular. For more realistic dialogues with patients, a user can use chatbots along with the deep learning models.

1.10.8 Personalized medical treatments

By evaluating patients' medical histories, symptoms, and tests, deep learning systems enable healthcare companies to provide individualized patient care. Natural language processing extracts suitable information from medical data for the health treatments.

1.10.9 Prescription audit

Deep learning algorithms compare treatments to patient health information for spotting, fixing potential investigative, or prescription faults.

1.10.10 Responding to patient queries

Chatbots based on deep learning assist healthcare professionals in identifying trends in patient symptoms.

Conclusion

One of the most powerful technologies for medical picture analysis is deep learning. We covered the basics of deep learning that relates to predictive analysis, healthcare statistics, and objective of relating deep learning in healthcare. It is found to be a critical area that can be particularly significant for medical imaging by observing at current advancements in deep learning. Deep learning technologies are also expected to be applied in the health industry's variable domain. Deep learning is frequently used in medical image processing, including ophthalmology, neuroimaging, ultrasonography, and other fields. Deep learning will be used in a growing number of medical sectors as the technology advances, and future deep learning will not just focus on neuroimaging but also on other aspects of bioinformatics and genetics.

In personalized medicine, deep learning is used to make health references and ailment actions centered on a person's hereditary genealogy, medical history, food, past ailments, stress levels, and other factors. As a result, deep learning has a significant effect on healthcare statistics processing. Healthcare's future has never been brighter. In healthcare, it can be extremely useful in assisting doctors, revolutionizing patient care, not just because AI and machine learning allow for the creation of solutions that are tailored to highly specific industry needs. The combination of 5G technology and deep learning technologies allows the machine to develop true intelligence. Simultaneously, the continual development of medical robots and intelligent medical devices supports deep learning applications at the hardware level, which greatly improves clinical outcomes.

References

[1] Anandhavalli Muniasamy, Deep Learning for Predictive Analytics in Healthcare, International Conference on Advanced Machine Learning Technologies and Applications,2019, pp 32-42.

[2]   Changwan Lee, Yeesuk Kim, Young-Soo Kim and Jongseong Jang, "Automatic Disease Annotation from Radiology Reports using Artificial Intelligence Implemented by a Recurrent Neural Network", American Journal of Roentgenology, 2019, Volume 212(4).

[3]   Hongming Li, Yong Fan, Brain decoding from functional MRI using long short term memory recurrent neural networks, Computer vision and Pattern recognition, 2018.

[4]   Lipton, Z.C., Kale, D.C., Elkan, C., Wetzel, R.: Learning to diagnose with LSTM recurrent neural networks. In: International Conference on Learning Representations, 2016.

[5]   Dr.N. Gnanasankaran and Dr.E. Ramaraj, IoT Sensor Networks with 5G Enabled Faster RCNN Based Generative Adversarial Network Model for Face Sketch Synthesis" in "Artificial Intelligence Techniques in IoT Sensor Networks", Taylor and Francis, CRC Press, 2021, 135-149.

[6]   B.Subashini and B.Sundaravadivazhagan, "A Review on Internet of Things (IoT): Security Challenges, Issuesand the Countermeasures approaches", Psychology and Education Journal 58(2),6544-6560.

[7]   Hongming, C., Engkvist, O., Wang, Y., Olivecrona, M., Blaschke, T.: The rise of deep learning in drug discovery. Drug Discov. Today 23(6), 1241–1250, 2018.

[8]   Gibson, E., et al.: NiftyNet: a deep-learning platform for medical imaging. Comput. Methods Programs Biomed. 158, 113–122, 2018.

[9]   Goodfellow, I., Bengio, Y., Courville, A.: Deep Learning, 1st edn. The MIT Press, Cambridge, 2016.

[10] Ajronline (n.d.). Retrieved from https://www.ajronline.org/doi/10.2214/AJR.18.19869

[11] Research.aimultiple (n.d.). Retrieved from https://research.aimultiple.com/deep-learning-in-healthcare

[12] Allerin (n.d.). Retrieved from https://www.allerin.com/blog/top-5-applications-of-deep-learning-in-healthcare

[13] data-flair. training (n.d.). Retrieved from https://data-flair.training/blogs/machine-learning-in-healthcare

[14] Hinton et all, A fast learning algorithm for deep belief nets, Neural Computation, 2006.

[15] Mingyukim, jihye hun et all, Deep learning in medical imaging, Neurospine 2019

[16] Hongming chin, Ola Engkvist et all, The rise of deep learning in drug discovery, drug discovery today, 2018.

[17] Andre Esteva, Alexandre Robicquet, Bharath Ramsundar et all, A guide to deep learning in healthcare, *Nature Medicine* volume 25, pages24–29 (2019)

[18] LeCun, Y., Bengio, Y. & Hinton, G. Deep learning. Nature 521, 436–444 (2015)

[19] Hirschberg, J. & Manning, C. D. Advances in natural language processing. Science 349, 261–266 2015).

[20] Litjens, G. et al. A survey on deep learning in medical image analysis. Med. Image Anal. 42, 60–88 (2017).

[21] Murphy KP. Machine learning: a probabilistic perspective. Cambridge (MA): MIT Press; 2012.

2

Deep Neural Network Architecture and Applications in Healthcare

Sarwath Unnisa[1], A. Vijayalakshmi[2], and Zainab Toyin Jagun[3]

[1]Research Scholar, Department of Computer Science,
CHRIST (Deemed to be University), India;
Email: Sarwath.unnisa@res.christuniversity.in
[2]Associate Professor, Department of Computer Science,
CHRIST (Deemed to be University), India;
Email: vijayalakshmi.nair@christuniversity.in
[3]Doctor of Philosophy (Real Estate), Faculty of Built Environment and
Surveying, Universiti Teknologi Malaysia, Malaysia;
Email: zainab.jagun@utm.my

Abstract

Gaining insights related to medical data has always been a challenge, as
limited technology delays treatment. Various types of data are collected
from the medical field, such as sensor data, that are heterogeneous in nature.
All of these are very poorly maintained and require more structuring. For
this reason, deep learning is becoming more and more popular in this
area. There are many challenges due to inadequate and irrelevant data.
Insufficient domain knowledge also adds to the challenge. Modern deep
learning models can help understand the dataset. This chapter provides
an overview of deep learning, its various architectures, and convolutional
neural networks. It also highlights how deep learning technologies can help
advance healthcare.

2.1 Introduction

Machine learning is a subset of artificial intelligence that can learn arbitrary correlations from data. The main purpose of machine learning is to derive a model. The representation of the model is based on how well the model fits and evaluates. For decades, machine learning has been used to transform raw data to recognize patterns. Deep learning is different from machine learning, but it is part of it. Since feature selection is not automated in machine learning, deep learning has been trending as it can automatically learn the feature. Therefore, learning and training are obtained in one shot [1], [2].

Healthcare is evolving as biomedical data plays an important role. For example, the correct data should lead to the correct treatment of the patient. Therefore, many aspects of responsibility for providing treatment are highlighted. On the other hand, the large amount of biomedical data presents many challenges. Therefore, it is very important to investigate the various relationships between these challenges. One such challenge is developing reliable tools in the medical field that can be data driven and deliver correct predictions [3]. For this purpose, deep learning in medicine can be used carefully. These approaches can be applied to genomics, wearable healthcare, and clinical image systems [3].

There are two types of deep learning network called as recursive neural network (RvNN) and recurrent neural network (RNN). A recurrent neural network is usually utilized for sequence processing such as language model; for example, techniques such as the hidden Markov model and n-gram model. The recurrent neural network is suggested so that the unrestrained history data can be used to obtain the hidden states so that the network can run for a longer time. The network will have three layers such as hidden layer, input layer, and output layer [4].

Recursive neural networks achieve predictions in hierarchical form and these are for objects such as graphs and trees. This network is usually trained using backpropagation technique [4], [5]. There are many deep learning networks, one of which is a convolutional neural network (CNN).

The main goal of the CNN network is to extract important data from the function and reduce the time required for extraction. To this end, CNNs use local receptive fields, weight sharing, and time and space subsampling. The main advantage of CNN is that it can learn individually from the training data. It also uses the subsampling feature to avoid time complexity and provide robustness [6].

CNN has two main methods: convolution and sampling. The convolution process has a trainable filter that can be used to de-convolve the input image and perform feature mapping to retrieve features. Then the bias is added and finally the convolution layer is created. The second process is the sampling process. This involves performing pooling steps, adding weights, and converting them to pixels. It also adds a bias to eventually generate an activation function, leading to feature growth from the feature map [6].

In this chapter, the importance of deep learning in healthcare, algorithms for deep learning neural networks, various applications of deep learning in healthcare, and how deep learning achieves predictions in the detection of pneumonia are discussed.

2.2 Deep Neural Network

Deep learning is considered a part of machine learning. Machine learning includes many steps such as pre-processing, feature extraction, dataset selection, and classification; however, all of this is not automated in machine learning. This is where deep learning comes into play. On the other hand, deep learning automates the training and learning, making the whole process easy and reliable [1], [2].

Artificial intelligence includes both deep learning and machine learning. Artificial intelligence is a platform that can infer, recognize, and adapt to new changes. The main difference between machine learning and deep learning is that the former improves performance when exposed to long-term data, while the latter uses a multi-layer neural network to learn from large amounts of data. Deep learning can be used without human help [7], [8]. Deep learning is a universal learning method because it can be used in all areas. It is robust because you can observe the functionality in a computerized way. Deep learning is scalable because its architecture can be used on a supercomputing scale. You can also use the same technique for different data types. This is also known as transfer learning and is very common [9].

Deep learning falls into four categories: supervised learning, semi-supervised learning, unsupervised learning, and reinforcement learning. Deep supervised learning is a subset of deep learning that uses labeled data. Many networks, such as recurrent neural networks (RNNs), convolutional neural networks (CNNs), and deep neural networks (DNNs), use this learning technique. The main advantage of this is that the information is obtained early. The disadvantage is that the technique is simple when compared to other techniques [6]. Deep semi-supervised technique uses the semi-labeled dataset. Generative adversarial networks (GANs) use this learning approach. The main advantage of this is that labeled data usage is minimized. The disadvantage is that sometimes irrelevant data can produce incorrect results [6]. Deep unsupervised learning can be executed with limited labeled data. Techniques such as generative networks, dimensionality reduction, and clustering use this technique. The disadvantage is that the results often tend to be unreliable [10]. Deep reinforcement learning relies on the communication with the situation rather than the sampled data. It was developed by Google DeepMind. Compared to traditional supervised learning, this learning is far more problematic [11].

One of the commonly used deep learning techniques is CNN. A CNN is similar to a multi-layer perceptron (MLP). There are neurons in MLP, and each neuron will consist of an activation function. CNN uses MLP with an arrangement that enables both rotation and translation to be used.

CNN has three basic layers which are convolutional layer, pooling layer, and fully connected layer [12]. This was developed by Hubel and Wiesel in 1960. CNN is a well-organized recognition algorithm which can be applied for identifying patterns and image processing techniques [13]. Typical CNN architecture is a fully connected network that has hidden activation layer, inputs, and weights [14]. Convolution layer is basic and important layer. The main work of this layer is to obtain pixel matrix for the image. Then an activation map is obtained for the image. The main function of the activation map is to save the features and reduce the amount of information that needs to be processed. The second layer is the pooling layer, which is typically employed to lower the map's dimension. This indicates that, of all the characteristics produced, only important features will get selected to reduce the invariance. Pooling can be classified into several categories, such as maximum pooling, average pooling, stochastic pooling, and spatial pyramid pooling. The best one is max pooling which will take biggest value from each matrix and make separate matrix. This will help to reduce the significant features of images. The last layer is the fully connected layer which is fed to the neural network. Usually, matrices are compressed before it is sent to the neurons. It is challenging to check the data after this point due to the existence of many hidden layers [15].

Deep learning has many architecture and the best ones among them have been listed below along with the year, developer's details, layers, and unique features.

Table 2.1 Architecture and its details

Sl. no.	Architecture	Year	Developed by	Layers	Unique features
1	LeNet [16]	1998	Yann LeCun, Leon Bottou, Yoshua Bengio, and Patrick	It has only two convolution and three fully connected layers.	• 60,000 parameters • The main uniqueness of this architecture from the rest is that this acts as a standard template which has activation function, pooling layer, and the fully connected layer at the end
2	AlexNet [17]	2012	Alex Krizhevsky, Ilya Sutskever, and Geoffrey Hinton	Eight layers	• 60 million parameters • ReLU activation function • Drop out feature
3	VGG16 [18]	2014	Karen Simonyan and Andrew Zisserman	16 layers comprising 13 convolutional and three fully connected layers	• 138 million parameters • This design is unique which means it is twice as deep as AlexNet, and they did this by stacking uniform convolutions • These also carry the ReLU technique from AlexNet. But these use small-sized filters of 2×2 and 3×3 matrices

Table 2.1 Continued

Sl. no.	Architecture	Year	Developed by	Layers	Unique features
4	Inception V1 [19]	2015	Christian Szegedy, Wei Liu, and Yangqing Jia	22 layer deep architecture	• The concept of a network in network approach, called inception • Uses parallel tower for concatenation • Removes computational bottleneck
5	Inception V3 [20]	2016	Christian Szegedy, Vincent Vanhoucke, Sergey Ioffe, JonathonShlens, and Zbigniew Wojna	48 layers	• The successor of Inception V1 • 24M parameters • Batch normalization technique. • Further reduces bottleneck problems better than Inception V1.
6	ResNet50 [21]	2015	Kaiming He, Xiangyu Zhang, and Shaoqing Ren	50 layers	• 26M parameters • Batch normalization • Used to design deeper networks

Table 2.1   Continued

Sl. no.	Architecture	Year	Developed by	Layers	Unique features
7	Xception [22]	2017	François Chollet	71 layers	• Arrived from inception • 23M parameters • Xception has arrived from the concept of inception. The difference is that the inception modules have been replaced by depth-wise separable convolutions
8	Inception V4 [23]	2014	Christian Szegedy, Sergey Ioffe, Vincent Vanhoucke, and Alex Alemi	22 layers	• 43M parameter. • Improvement of Inception V3. • Stem module (uniform blocks for the inception module)
9	Inception-ResNet-V2 [23]	2014	Christian Szegedy, Sergey Ioffe, Vincent Vanhoucke, and Alex Alemi	164 layers	• 56M parameters • Improvised version of Inception V3 • The difference being that the inception modules are converted into residual inception blocks

Table 2.1   Continued

Sl. no.	Architecture	Year	Developed by	Layers	Unique features
10	RESNETXT50 [24]	2017	Saining Xie, Ross Girshick, and Piotr Dollár	50 layers	• 25M parameters • The difference between ResNet and this architecture is that parallel towers have been added between each module • There are 32 additional towers which are added

When deciding on the best architecture for the deep learning model, it's best to thoroughly explore the benefits of the architecture. The following conclusions were drawn from this study: the higher the number of convolutions and the denser the layers, the better the results. In addition, dropout capabilities and batch normalization techniques improve network performance. More reliable approaches are AlexNet, ResNet, VGG Net, and Inception Net with error rates of 15.3%, 3.6%, 7.3%, and 6.6%, respectively [17], [18], [21], [23].

2.3 Deep Learning Architectures Applied in the Healthcare Field

2.3.1 *Alzheimer's disease*

Alzheimer's is an irreversible neurodegenerative disorder. The disease has different treatments that can control the severity of the disease but cannot eradicate it. Therefore, early diagnosis is essential, which can be achieved through profound learning. For this, many researches have come to analyze for early diagnosis. Studies also claim that the genetic makeup of an individual is responsible for the traits of Alzheimer's disease, and, hence, the evaluation of genes is vital. CNN is used to make this possible by detecting

Alzheimer's at an early stage by identifying brain patterns that only occur in these patients [25].

2.3.2 Brain mris

Deep learning algorithms are used in training 3D medical images that can detect various diseases of the human brain with associated parameters [3].

2.3.3 Osteoarthritis

Deep learning algorithms like CNN can classify segmented MRI images of the tibia cartilage of the knee into osteoarthritis and non-osteoarthritis which can help the doctors for quick analysis and further treatment [26].

2.3.4 Breast cancer

Deep learning is used to diagnose between cancer and non-cancer images with precision in contrast to errors that may have occurred due to incorrect processing results. Other techniques used to detect breast cancer are typically vague and less robust. Numerous techniques and architectures have been implemented in deep learning, such as the automatic encoder architecture which is noise tolerant [27].

2.3.5 Diabetic retinopathy

Deep learning is used to develop an algorithm that detects diabetic retinopathy and diabetic macular edema in patients with the bottom of the retina. For this purpose, a neural network with image classification skills is trained in such a way that it can classify diabetic retinopathy more efficiently and precisely than trained ophthalmologists [28].

2.3.6 Forecasting type of medicine based on patient history

Deep learning is used to give personalized medicines to patients with past illness as healthcare field has a lot of dependencies and observations which are sometimes recorded through handwritten notes which are irregular at a time. Using deep learning to predict the type of drug a patient needs through medical history will be more effective and quicker [29].

2.3.7 Forecasting diseases through patient's clinical status

Deep learning holds the promise of advancing clinical research and better clinical decision-making. In this way, deep learning promises to derive the patient's diseases that also lead to better and rapid decision-making in the case of very serious diseases. That decision will be based on electronic health records. Many diseases that require rapid treatments such as diabetes and schizophrenia may also be detected [30].

2.3.8 Forecasting suicide

The health industry is prevalent and evolving rapidly. For this reason, quick action needs to be taken. With the help of deep learning and electronic medical records, many rapid actions can be considered. Such activity is predicted if the patient suffering from severe mental illness will die by suicide. That is done through the suicide risk assessment. Then the patients are grouped and the derived representation will show whether the grouping contains a risk or not. Once this scoring system is generated, it categorizes if the risk is moderate or severe. It is more effective than the risk evaluated by the clinicians [31].

2.3.9 Forecasting readmission of patients after the discharge

Deep learning has evolved a lot in the field of healthcare that it is able to detect from the regular clinical visits of the patients whether there is a possibility of readmission of the patients. For this purpose, the visit details are used [32].

2.3.10 Forecasting disease from lab test

Deep learning is used to predict illness from laboratory tests by longitudinal analysis of laboratory tests. By using long short-term memory recurrent neural network and CNN networks a predictive model can be generated which will be able to tell what disease the patient is suffering from by just analyzing the lab reports from previous health records [33].

2.3.11 Forecasting the quality of sleep by awake time activities

Sleep is extremely important for health. Less sleep can lead to stress in life, leading to serious health complications. Therefore, the awakening activities that we carry out during the day will strongly affect the duration and quality of sleep. Consequently, deep learning was used to determine the quality of sleep by waking-time activities. To do so, a handheld medical device is used. These predictive models were developed to determine sleep quality [34].

2.4 Pneumonia Detection using Deep Learning from X-ray Images

2.4.1 *Overview*

Pneumonia is a bacterial infection that affects the lungs. It can affect one or both lungs. It is known to be a deadly disease that can kill people [35]. It continues to be the cause of death around the world [36]. However, the disease can be easily dealt with if treated in time with the help of antibiotics and antivirals [37]. One of the best ways to treat pneumonia is early detection, which can be done with a chest X-ray [35]. Pneumonia is said to be a common disease that occurs regularly, especially in patients with a cardiovascular history or lung disease [36].

However, detecting pneumonia just by looking at X-rays can be a very confusing task, even for some professionals. X-ray images can be mistaken for another benign disease, which can lead to misdiagnosis, which is confusing for professionals. Therefore, there is a need for a more systematic and computational method to help physicians diagnose pneumonia from radiographic images [38]. Image classification is a technique that allows to classify images more systematically and accurately. CNN (convolutional neural network) is a branch of deep learning neural networks used for image recognition and classification. CNNs can extract features from images that are useful for classification. CNN has found a significant contribution in the field of image classification and also enables localization in computer vision [39].

Only few studies have focused on pneumonia and its detection techniques. The purpose of this study is to understand and identify the differences between radiographic images of healthy people and patients with pneumonia. In addition, identification should be completed as quickly as possible so that the diagnosis can be made and the patient can be treated. Pneumonia is a common infection caused by bacteria, viruses, or fungi attacking the lungs. The word pneumonia comes from the Greek word "pneumon", which means lungs. So when there is an inflammation in the lung's parenchyma, then this is associated with pneumonia. It can also be caused by exposure to certain chemicals or food aspiration [40].

The definition of pneumonia is different based on varied opinions. Pneumonia is defined as "acute lower respiratory tract disease" by the World Health Organization [41]. Pneumonia occurs when fluid fills the air sacs of the lungs and reduces the flow of carbon dioxide and oxygen between the blood and lungs. This reduces the amount of air that enters the lungs, resulting in problems with human breathing. Other symptoms include shortness of breath, fever, cough, and chest pain [42].

Older people over the age of 65 are at increased risk. Children under the age of 5 and patients with other respiratory and heart diseases are also at great risk [40], [42]. Risk factors associated with pneumonia include patients with a history of alcoholism, lung disease, heart disease, bronchial asthma, and patients receiving immunosuppressive therapy [36]. India has the world's largest number of child deaths from pneumonia, with approximately 300,000 children dying in 2016 [43].

There are several approaches used to diagnose pneumonia [44]. Other alternative approaches to identify pneumonia with X-rays include pulse oximetry, sputum and blood gas analysis, bronchoscopy, and complete blood count. There are three types of pneumonia: bacterial infections, viral infections [45], and fungal infections. Because pneumonia is based on the causative agent, bacterial pneumonia is cured using antibiotics, viral pneumonia is cured using antiviral drugs, and fungal pneumonia is cured using antiviral drugs.

Chest X-ray is an important tool in diagnosing pneumonia, and all subsequent decisions are based on this radiological finding. It is also very cheap compared to other methods that people can buy. When looking at a chest X-ray of pneumonia-suspected patient, the person responsible looks for white spots on the lungs called infiltrates that classify the contamination. These are also found in patients with tuberculosis and bronchitis [43]. Early detection of pneumonia is always beneficial, as delayed treatment can have a variety of consequences [46].

Deep learning can be used for this purpose. Deep learning is a subfield of machine learning that is part of artificial intelligence. Many biomedical health problems, such as cancer, can be detected using artificial-intelligence-based models. These models may be able to easily identify features that are hidden and not easily visible to the professional. Convolutional neural networks are excellent tools in the fields of machine learning and deep learning. This tool is most widely and recently used in healthcare to extract many new features and distinguish between classes such as benign and malignant cancers, infections, or healthy tumors [47].

ResNet50 is a convolutional neural network architecture that contains about 50 layers. The layers are the depth of these networks. The complete name of ResNet is called the residual network. This is a classic neural network that is the pillar of many computing tasks. This architecture produces a pre-trained version that has been tested with millions of photos; so any element can be categorized fairly quickly. These provide a solid combination of benefits and can also allow for faster training [48]. Therefore, this is considered to be the most efficient architecture. The main idea of this network is to get only the important features from the previous layer, called the residual features. This network allows you to train much deeper neural networks without problems [49].

2.4.2 Methodology

As mentioned, the deep residual features are extracted from the last output layer. The ResNet architecture has five convolution blocks stacked on top of each other. These stacks are different from traditional convolutional neural networks. This allows networks to be trained in a faster and more cost-effective way [50].

Figure 2.1 shows the methodology and abstract view of the working model. The first step is to pre-process the chest X-ray image after performing the augmentation process. The second step is the training process where the dataset is split into a training set and a test set. Approximately 80% of the data is available for training datasets. The second part of the training process is the application of deep learning models. Here, the ResNet50 model applies. The third phase uses the remaining 20% of the dataset to test the accuracy of the model and other performance metrics.

2.4.2.1 *Visualizing the images*
First, the dataset is uploaded to MATLAB as shown above and then the loaded dataset is visualized as shown in Figure 2.2. This will show that the uploaded test data is correctly classified. Visualizing the image helps you better understand how each class differs from the other. It will also determine the type of CNN architecture to use. Here, the ResNet50 architecture was used.

2.4.2.2 *Resizing*

This feature helps to resize the image used for training according to ResNet50, as training requires the same matrix dimensions. ResNet has the image size set to [224 224]; so all images will be converted as shown in the code.

```
% Resizing all training images to [224 224] for ResNet architecture
Auimds=augmentedImageDatastore([224  224],imdsTrain,'Data
Augmentation',augmenter);
% Data Augumentation
augmennter = imageDataAugmennter( ...
'RandRotation',[-6 6],'RandXReflection',1,...
'RandYReflection',1,'RandXShear',[-0.06  0.06],'RandYShear',[-0.06
0.06]);
```

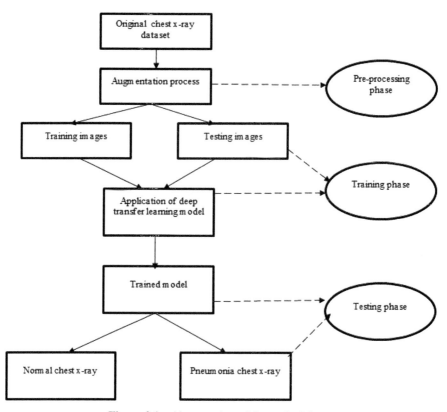

Figure 2.1   Abstract view of the methodology.

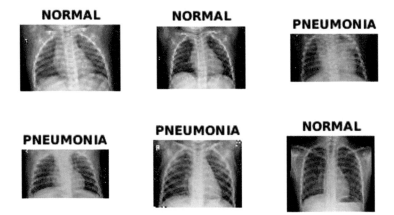

Figure 2.2 Visualizing the loaded dataset.

2.4.3 Results

2.4.3.1 *ROC curve*
The data is split into 10 folds for analysis. This type of validation will help to provide better estimate of the performance. ResNet50 architectures are adopted as it is considered to be very effective for medical imaging. Figure 2.3 indicates the ROC curve which will help to identify the AUC. Here, AUC (area under the curve) is approximately 95.9%.

2.4.3.2 *Confusion matrix*
Figure 2.4 shows the confusion matrix for a particular dataset. The confusion matrix tells about images that are correctly classified and those that are not. Approximately 47.9% of the data are correctly classified as healthy people without pneumonia, 2.1% are misclassified as healthy people with pneumonia, 2.1% are misclassified as healthy people despite having pneumonia, and 47.9.% are correctly classified as having pneumonia. Therefore, the ResNet50 architecture achieved an overall accuracy of 95.9%.

Conclusion

Innovation in technology has enormously contributed to the healthcare industry that supports an easy and comfortable lifestyle for humanity. Various algorithms in deep learning helped in easy and quick analysis of healthcare data accurately. This chapter discusses various deep learning algorithms with their architectures in detail. Applications of deep learning in various healthcare

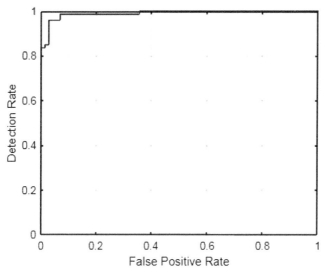

Figure 2.3   ROC curve.

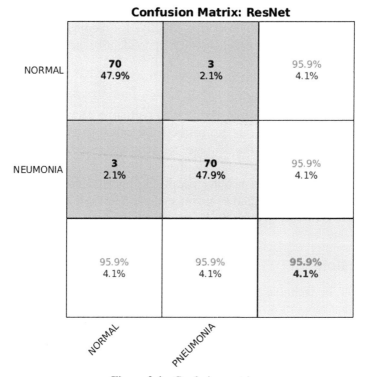

Figure 2.4   Confusion matrix.

domains are also listed that can contribute to understanding the importance of deep learning in health data analysis. Further, the ResNet algorithm is used in classification of pneumonia from X-ray images. The result shows a good accuracy of 95.9% in classifying X-rays with pneumonia.

References

[1] Y. Lecun, Y. Bengio, and G. Hinton, "Deep learning," *Nature*, vol. 521, no. 7553, pp. 436–444, 2015, doi: 10.1038/nature14539.

[2] Z. Zhang, P. Cui, and W. Zhu, "Deep Learning on Graphs: A Survey," *IEEE Trans. Knowl. Data Eng.*, vol. 14, no. 8, pp. 1–1, 2020, doi: 10.1109/tkde.2020.2981333.

[3] T. Brosch and R. Tam, "Manifold learning of brain MRIs by deep learning," *Lect. Notes Comput. Sci. (including Subser. Lect. Notes Artif. Intell. Lect. Notes Bioinformatics)*, vol. 8150 LNCS, no. PART 2, pp. 633–640, 2013, doi: 10.1007/978-3-642-40763-5_78.

[4] S. Liu, N. Yang, M. Li, and M. Zhou, "A recursive recurrent neural network for statistical machine translation," *52nd Annu. Meet. Assoc. Comput. Linguist. ACL 2014 - Proc. Conf.*, vol. 1, pp. 1491–1500, 2014, doi: 10.3115/v1/p14-1140.

[5] C. Goller and A. Kuechler, "Learning task-dependent distributed representations by backpropagation through structure," *IEEE Int. Conf. Neural Networks - Conf. Proc.*, vol. 1, pp. 347–352, 1996, doi: 10.1109/icnn.1996.548916.

[6] K. O'Shea and R. Nash, "An Introduction to Convolutional Neural Networks," pp. 1–11, 2015, [Online]. Available: http://arxiv.org/abs/1511.08458.

[7] J. De Fauw *et al.*, *Clinically applicable deep learning for diagnosis and referral in retinal disease*, vol. 24, no. 9. 2018.

[8] Topol EJ., "High-performance medicine: the convergence of human and artificial intelligence.," *Nat Med.*, vol. 25(1), pp. 44–56, 2019.

[9] B. Van Essen, H. Kim, R. Pearce, K. Boakye, and B. Chen, "Lbann," pp. 1–6, 2015, doi: 10.1145/2834892.2834897.

[10] J. Karhunen, T. Raiko, and K. H. Cho, "Unsupervised deep learning: A short review," *Adv. Indep. Compon. Anal. Learn. Mach.*, pp. 125–142, 2015, doi: 10.1016/B978-0-12-802806-3.00007-5.

[11] K. Arulkumaran, M. P. Deisenroth, M. Brundage, and A. A. Bharath, "Deep reinforcement learning: A brief survey," *IEEE Signal Process. Mag.*, vol. 34, no. 6, pp. 26–38, 2017, doi: 10.1109/MSP.2017.2743240.

[12] U. R. Acharya *et al.*, "A deep convolutional neural network model to classify heartbeats," *Comput. Biol. Med.*, vol. 89, no. September, pp. 389–396, 2017, doi: 10.1016/j.compbiomed.2017.08.022.

[13] T. Liu, S. Fang, Y. Zhao, P. Wang, and J. Zhang, "Implementation of Training Convolutional Neural Networks," 2015, [Online]. Available: http://arxiv.org/abs/1506.01195.

[14] T. N. Sainath, A. R. Mohamed, B. Kingsbury, and B. Ramabhadran, "Deep convolutional neural networks for LVCSR," *ICASSP, IEEE Int. Conf. Acoust. Speech Signal Process. - Proc.*, pp. 8614–8618, 2013, doi: 10.1109/ICASSP.2013.6639347.

[15] A. Ajit, K. Acharya, and A. Samanta, "A Review of Convolutional Neural Networks," *Int. Conf. Emerg. Trends Inf. Technol. Eng. ic-ETITE 2020*, no. October, 2020, doi: 10.1109/ic-ETITE47903.2020.049.

[16] Y. Lecun, L. Bottou, Y. Bengio, and P. Ha, "LeNet," *Proc. IEEE*, no. November, pp. 1–46, 1998.

[17] A. Krizhevsky and G. Hinton, "ImageNet Classification with Deep Convolutional Neural Networks," *ImageNet Large Scale Vis. Recognit. Chall. 2012*, p. 27, 2012, [Online]. Available: http://www.image-net. org/challenges/LSVRC/2012/supervision.pdf%0Ahttp://image-net.org/ challenges/LSVRC/2012/results.html#t1.

[18] K. Simonyan and A. Zisserman, "Very deep convolutional networks for large-scale image recognition," *3rd Int. Conf. Learn. Represent. ICLR 2015 - Conf. Track Proc.*, pp. 1–14, 2015.

[19] C. Szegedy *et al.*, "Going deeper with convolutions," *Proc. IEEE Comput. Soc. Conf. Comput. Vis. Pattern Recognit.*, vol. 07-12-June, no. October, pp. 1–9, 2015, doi: 10.1109/CVPR.2015.7298594.

[20] C. Szegedy, V. Vanhoucke, S. Ioffe, J. Shlens, and Z. Wojna, "Rethinking the Inception Architecture for Computer Vision," *Proc. IEEE Comput. Soc. Conf. Comput. Vis. Pattern Recognit.*, vol. 2016-Decem, pp. 2818–2826, 2016, doi: 10.1109/CVPR.2016.308.

[21] K. He, X. Zhang, S. Ren, and J. Sun, "Deep residual learning for image recognition," *Proc. IEEE Comput. Soc. Conf. Comput. Vis. Pattern Recognit.*, vol. 2016-Decem, pp. 770–778, 2016, doi: 10.1109/CVPR.2016.90.

[22] F. Chollet, "Xception: Deep learning with depthwise separable convolutions," *Proc. - 30th IEEE Conf. Comput. Vis. Pattern Recognition, CVPR 2017*, vol. 2017-Janua, pp. 1800–1807, 2017, doi: 10.1109/CVPR.2017.195.

[23] C. Szegedy, S. Ioffe, V. Vanhoucke, and A. A. Alemi, "Inception-v4, inception-ResNet and the impact of residual connections on learning," *31st AAAI Conf. Artif. Intell. AAAI 2017*, pp. 4278–4284, 2017.

[24] S. Xie, R. Girshick, and P. Doll, "Aggregated Residual Transformations for Deep Neural Networks," *Cvpr*, pp. 1492–1500, 2017.

[25] D. Pan, Y. Huang, A. Zeng, L. Jia, and X. Song, *Early Diagnosis of Alzheimer's Disease Based on Deep Learning and GWAS*, vol. 1072, no. May 2020. Springer Singapore, 2019.

[26] A. Prasoon, K. Petersen, C. Igel, F. Lauze, E. Dam, and M. Nielsen, "Deep feature learning for knee cartilage segmentation using a triplanar convolutional neural network," *Lect. Notes Comput. Sci. (including Subser. Lect. Notes Artif. Intell. Lect. Notes Bioinformatics)*, vol. 8150 LNCS, no. PART 2, pp. 246–253, 2013, doi: 10.1007/978-3-642-40763-5_31.

[27] "Cheng, JZ, Ni, D, Chou, YH, Qin, J, Tiu, CM, Chang, YC, Huang, CS, Shen, D & Chen, CM 2016, 'Computer-Aided Diagnosis with Deep Learning Architecture: Applications to Breast Lesions in US Images and Pulmonary Nodules in CT Scans', Scientific reports, vol."

[28] V. Gulshan *et al.*, "Development and validation of a deep learning algorithm for detection of diabetic retinopathy in retinal fundus photographs," *JAMA - J. Am. Med. Assoc.*, vol. 316, no. 22, pp. 2402–2410, 2016, doi: 10.1001/jama.2016.17216.

[29] T. Pham, T. Tran, D. Phung, and S. Venkatesh, "DeepCare: A deep dynamic memory model for predictive medicine," *Lect. Notes Comput. Sci. (including Subser. Lect. Notes Artif. Intell. Lect. Notes Bioinformatics)*, vol. 9652 LNAI, no. i, pp. 30–41, 2016, doi: 10.1007/978-3-319-31750-2_3.

[30] R. Miotto, L. Li, B. A. Kidd, and J. T. Dudley, "Deep Patient: An Unsupervised Representation to Predict the Future of Patients from the Electronic Health Records," *Sci. Rep.*, vol. 6, no. January, pp. 1–10, 2016, doi: 10.1038/srep26094.

[31] T. Tran, T. D. Nguyen, D. Phung, and S. Venkatesh, "Learning vector representation of medical objects via EMR-driven nonnegative restricted Boltzmann machines (eNRBM)," *J. Biomed. Inform.*, vol. 54, pp. 96–105, 2015, doi: 10.1016/j.jbi.2015.01.012.

[32] P. Nguyen, T. Tran, N. Wickramasinghe, and S. Venkatesh, "Deepr: A Convolutional Net for Medical Records," *IEEE J. Biomed. Heal. Informatics*, vol. 21, no. 1, pp. 22–30, 2017, doi: 10.1109/JBHI.2016.2633963.

[33] N. Razavian, J. Marcus, and D. Sontag, "Multi-task Prediction of Disease Onsets from Longitudinal Lab Tests," vol. 56, 2016, [Online]. Available: http://arxiv.org/abs/1608.00647.

[34] A. Sathyanarayana *et al.*, "Sleep quality prediction from wearable data using deep learning," *JMIR mHealth uHealth*, vol. 4, no. 4, pp. 1–13, 2016, doi: 10.2196/mhealth.6562.

[35] P. Rajpurkar *et al.*, "CheXNet: Radiologist-level pneumonia detection on chest X-rays with deep learning," *arXiv*, pp. 1–22, 2017.

[36] I. Koivula, M. Sten, and P. H. Makela, "Risk factors for pneumonia in the elderly," *Am. J. Med.*, vol. 96, no. 4, pp. 313–320, 1994, doi: 10.1016/0002-9343(94)90060-4.

[37] E. Ayan and H. M. Ünver, "Diagnosis of pneumonia from chest X-ray images using deep learning," *2019 Sci. Meet. Electr. Biomed. Eng. Comput. Sci. EBBT 2019*, pp. 0–4, 2019, doi: 10.1109/EBBT.2019.8741582.

[38] K. El Asnaoui, Y. Chawki, and A. Idri, "Automated Methods for Detection and Classification Pneumonia based on X-Ray Images Using Deep Learning," *arXiv*, 2020.

[39] J. Merkow, "Pneumonia detection in chest radiographs," *arXiv*, 2018.

[40] M. Ozsoz, "Viral and Bacterial Pneumonia Detection using Arti cial Intelligence in the Era of COVID-19," pp. 1–20.

[41] N. America, N. America, and K. McIntosh, "COMMUNITY -A CQUIRED P NEUMONIA IN CHILDREN," *N. Engl. J. Med.*, vol. 346, no. 6, pp. 429–437, 2002, [Online]. Available: http://www.ncbi.nlm.nih.gov/pubmed/11832532.

[42] S. Shah, H. Mehta, and P. Sonawane, Pneumonia detection using convolutional neural networks, no. April. Springer Singapore, 2020.

[43] A. Sharma, D. Raju, and S. Ranjan, "Detection of pneumonia clouds in chest X-ray using image processing approach," *2017 Nirma Univ. Int. Conf. Eng. NUiCONE 2017*, vol. 2018-Janua, pp. 1–4, 2018, doi: 10.1109/NUICONE.2017.8325607.

[44] J. Chastre and J. Fagon, "State of the Art Ventilator-associated Pneumonia," *Am J Respir Crit Care Med*, vol. 165, no. 23, pp. 867–903, 2002, doi: 10.1164/rccm.2105078.

[45] O. Ruuskanen, E. Lahti, L. C. Jennings, and D. R. Murdoch, "Viral pneumonia," *Lancet*, vol. 377, no. 9773, pp. 1264–1275, 2011, doi: 10.1016/S0140-6736(10)61459-6.

[46] D. Varshni, K. Thakral, L. Agarwal, R. Nijhawan, and A. Mittal, "Pneumonia Detection Using CNN based Feature Extraction," *Proc. 2019 3rd IEEE Int. Conf. Electr. Comput. Commun. Technol. ICECCT 2019*, pp. 1–7, 2019, doi: 10.1109/ICECCT.2019.8869364.

[47] M. F. Hashmi, S. Katiyar, A. G. Keskar, N. D. Bokde, and Z. W. Geem, "Efficient pneumonia detection in chest xray images using deep transfer learning," *Diagnostics*, vol. 10, no. 6, pp. 1–23, 2020, doi: 10.3390/diagnostics10060417.

[48] M. Farooq and A. Hafeez, "COVID-ResNet: A Deep Learning Framework for Screening of COVID19 from Radiographs," 2020, [Online]. Available: http://arxiv.org/abs/2003.14395.

[49] M. M. Eid and Y. H. Elawady, "Efficient Pneumonia Detection for Chest Radiography Using ResNet-Based SVM," *Eur. J. Electr. Eng. Comput. Sci.*, vol. 5, no. 1, pp. 1–8, 2021, doi: 10.24018/ejece.2021.5.1.268.

[50] A. Mahmood *et al.*, "Automatic hierarchical classification of kelps using deep residual features," *Sensors (Switzerland)*, vol. 20, no. 2, pp. 1–20, 2020, doi: 10.3390/s20020447.

3

The State of the Art of using Artificial Intelligence for Disease Identification and Diagnosis in Healthcare

Iman El Mir[1] and Said El Kafhali[2]

[1]Hassan First University of Settat, Institute of Sports Sciences, Computer, Networks, Modeling, and Mobility Laboratory (IR2M), Morocco;
Email: iman.elmir@uhp.ac.ma
[2]Hassan First University of Settat, Faculty of Sciences and Techniques, Computer, Networks, Modeling, and Mobility Laboratory (IR2M), Morocco;
Email: said.elkafhali@uhp.ac.ma

Abstract

Healthcare is a multidisciplinary term, which refers to a system that involves the development of health services to satisfy people's medical needs. Over the past years, healthcare analysis processed various types of diseases including diabetic retinopathy, cancers, Alzheimer's, cardiovascular diseases, vascular dementia, chronic disease, osteoporosis, heart disease, strokes, and epilepsy, to mention just a few, using artificial intelligence. Nowadays, artificial intelligence proposes various solutions in several fields such as marketing and profit analysis, financial services, healthcare, and others. By applying artificial intelligence in healthcare, machines can be a lifesaver. In healthcare, machines play an essential role in disease identification and diagnosis. Artificial intelligence offers many solutions to healthcare using deep learning, artificial neural networks, and machine learning techniques. In this chapter, we review papers that utilize artificial intelligence in healthcare applications to provide better solutions. Furthermore, we concentrate on which type of artificial intelligence techniques we can use for which type of problems people usually face in the healthcare field. Readers and investigators can benefit from this chapter by understanding the applications, roles, challenges, and future perspectives of artificial intelligence for healthcare services, especially for disease prediction and diagnosis.

3.1 Introduction

With the rapid development of the healthcare system, new technologies and Internet-based health services are invented and improved. Progress in the healthcare system, as well as other fields, is inseparable from the development of information science and communication technology. Accurate and safe information has become a means of communication between organizations and even health companies in today's society. It is crucial to ensure the security of the transmission and processing of different health information. Today, we need to find a new way to ensure the health services of our health system. Healthcare has recently seen important advances and is considered as another revolutionary and significant scientific progress. Healthcare is one of the fields of rapid development and is today at the core of complete transformation and evolution. From now on, doctors and medical personnel will have to utilize new approaches, such as machine learning [1], [2], [3], [4], [5], [6], [7], [8], [9], artificial intelligence [10], big data [11], data mining [12], cloud, and Internet of Things (IoT) [13], to perform all patient care, including diagnosis of the disease, early detection of the disease, and prediction of disease progression and progression.

Artificial intelligence dates back to the 1950s as a new area of research; nowadays, it is gradually becoming an intelligent system that has revolutionized industries in various fields [14]. The purpose of artificial intelligence is to construct a novel algorithm that allows machines to operate intelligently without human intervention. The main basis of artificial intelligence is machine learning, which allows enhancing their performance and skills by constantly analyzing their reactions to the real world. Artificial intelligence evenly refers to cases in which machines can simulate and reproduce human

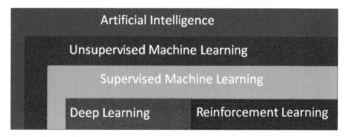

Figure 3.1   Summary of artificial intelligence techniques.

bits of intelligence in learning and analysis and, consequently, can solve complex problems such as NP-complete problems [15].

With advances in artificial intelligence, machine learning has become the key technique for representing and processing data in several fields. For healthcare services, the main problem is to efficiently analyze and retrieve independent patient data from a large volume of heterogeneous data over a long period. Machine learning is an advance in artificial intelligence in computer technology that allows machines to learn in the absence of having been explicitly programmed for this goal. Big data is the motor of machine learning and is a very important way that computers learn, analyze, and train. Machine learning is very effective in contexts where information needs to be discovered from large, diverse, and changing datasets, such as health data. Analyzing such data is much more efficient than traditional techniques in terms of precision and speed. For instance, based on information corresponding to a disease, such as prediction and diagnosis, it can be done within a millisecond. Therefore, it is significantly more helpful than traditional methods for different health services. However, artificial intelligence habitually refers to a machine (or computer) that learns from raw data with a certain degree of autonomy, as it happens with unsupervised machine learning, supervised machine learning, deep learning, and reinforcement learning as shown in Figure 3.1 [16].

With unsupervised machine learning, computing machines learn to map between input data and the prescribed output categories using homemade extractors conceived by human experts in this area. On the other hand, with supervised machine learning, the computing machines develop their output data as stated by the distribution and structures of the input data. In the case of deep learning, the computing machines constitute data by adjusting weights through a layered network of neurons. Finally, with reinforcement learning, the computing machines select the actions that produce the highest probability of a specified result.

Besides, nowadays, researchers are more concerned with developing, testing, implementing, and validating machine learning algorithms for healthcare to improve both the efficiency and quality of personal care [17]. The machine learning model [2] is an algorithm that learns from data to perform a task or make a decision and has explicitly programmed behavior. Machine learning methods can be used with certain other technologies such as fog/edge computing [18], the Internet of Things, and the cloud to render better solutions and serve as a lifesaver in healthcare [19], [20].

With rapidly improving computational power and the accessibility of big data, deep learning has become the default artificial intelligence approach that is involved since it can learn much more than classical machine learning methods. This new technique considerably simplifies the features of the engineering process and is applied to raw data immediately, based on the construction of artificial neural networks. Therefore, this allows more researchers to operate new ideas faster and easier in the field of healthcare. However, this chapter focuses on how to develop and apply artificial intelligence methods to healthcare. The main contributions of this chapter are the following.

- We begin by reviewing some applications of artificial intelligence and machine learning in healthcare, specifically, for disease prediction and diagnosis.
- We present the steps to develop machine learning models for healthcare and a brief overview of the challenges faced when deploying these models in the healthcare industry.
- We illustrate some challenges in developing artificial intelligence models for healthcare care.
- We conclude by providing certain solutions on how to address these models from the directions of different clinical needs.

The remainder of this chapter is as follows. A rapid review of the literature on machine learning and artificial intelligence in healthcare is presented in Section 3.2. Section 3.3 provides the steps to develop machine learning algorithms for healthcare. Some disease predictions and diagnoses using artificial intelligence are presented in Section 3.4. Section 3.5 provides some challenges in the development of artificial intelligence models for healthcare. Finally, the final remarks are presented in Section 3.6.

3.2 A Review of the Literature on Machine Learning and Artificial Intelligence in Healthcare

3.2.1 *Machine learning applications in healthcare*

The machine learning technique plays a significant role in healthcare. It is currently being developed and implemented for use in various targeted health applications, such as patient monitoring, medical diagnostics, health systems learning, and clinical decision support [21]. These englobe a family of methods including diverse concepts like supervised learning, cognitive learning, unsupervised learning, deep learning, and reinforcement learning-based algorithms that can be used to analyze and interpret complex healthcare data in situations where classical statistical algorithms may not be able to perform. The hopeful avenues for machine learning algorithms in healthcare are the development of automated risk prediction methods that can be used to guide clinical care. The purpose of a predictive model is determined by the heterogeneity of the data, depth of the data, breadth of data, modeling task, the choice of machine learning choice, and the selection techniques [22]. Even if machine learning algorithms have often been used in healthcare, a greater number of jurisdictions do not allow these algorithms to be the ultimate decision-maker; they are mostly used as a detection method or as a diagnostic aid.

The use of machine learning for health services is not novel, and it has known an increasing expansion in both the public and private health sectors. The most significant machine learning services in healthcare are illustrated in Figure 3.2. First, there are the identification and diagnosis, which involves identifying the patients' disease, monitoring their condition, and suggesting the necessary measures to prevent it. There is also drug discovery and manufacturing, which includes automating the identification of a new drug or lead molecule to minimize errors and enhance efficiency in drug development and discovery. Thanks to deep learning and computer vision, medical imaging is also a very important application in healthcare, which facilitates the detection of microscopic abnormalities; therefore, doctors can suggest an appropriate diagnosis. Another application is patient treatment and monitoring, which encompasses the planning and management of specific actions, based on the massive amounts of datasets and electronic health records presented, to make patients healthy. Smart health records that allow patients to directly share their data is another application where machine learning has been used to save effort, time, and cost. Finally, many machine learning models, such as artificial neural networks, are used in disease prediction, which helps predict not only minor illnesses but also

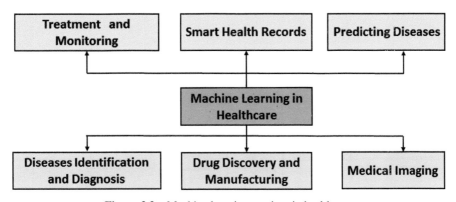

Figure 3.2 Machine learning services in healthcare.

severe infectious ones based on a considerable amount of data possessed from social media platforms, satellites, websites, etc.

Currently, the machine learning algorithm is being used in one of the most serious unprecedented public health crises in the world, the new coronavirus epidemic (SARS-CoV-2 or COVID-19). It has helped scientists classify COVID-19 genomes [23], identify potential drug candidates against COVID-19 [24], predict COVID-19-related protein structures for vaccine formulation [25], and in other useful applications for the COVID-19 virus [26]. The authors of [23] identify an intrinsic SARS-CoV-2 genomic signature and use it with a machine learning algorithm that combines various techniques for extremely accurate classification of the entire SARS-CoV-2 virus genomes. The proposed algorithm allows for a 100% accurate classification of SARS-CoV-2 virus sequences and finds the best suitable relationships between more than 5000 viral genomes in a few minutes. However, the algorithm can render a good real-time possibility for taxonomic classification. Ge *et al.* [24] proposed a framework that applies statistical analysis and machine learning techniques to incorporate and utilize large-scale knowledge graphs, literature, and transcriptomic data to find prospective drug candidates for the COVID-19 virus. They used past MERS-CoV and SARS-CoV data to demonstrate that the proposed framework can predict helpful drug candidates against a certain coronavirus. Khan *et al.* [27] suggested a deep learning model named CoroNet to identify COVID-19 infected cases by the usage of radiography images based on a deep convolutional neural network. The proposed model has been tested on the dataset used and the results showed that the model achieved an overall accuracy of 89.6%.

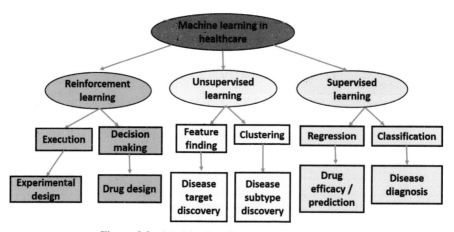

Figure 3.3   Machine learning services in healthcare.

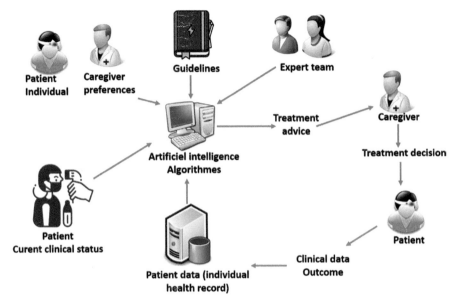

Figure 3.4 A typical architecture of artificial intelligence applications in healthcare.

Figure 3.3 shows which type of machine learning techniques we can use for which type of problem in the healthcare field.

3.2.2 Applications of artificial intelligence in healthcare

Artificial intelligence introduces a large number of intelligent algorithms that can be used in Industry 4.0. The majority of these algorithms have an important effect on the healthcare field. The performance of artificial intelligence techniques in healthcare is conditional on its key parameters and the quality of the health data. A typical architecture of artificial intelligence applications in healthcare, with great possibility of reducing healthcare costs while preserving healthcare quality, is the transformation of care provided only by professionals into personalized patient care with shared responsibilities and self-empowerment (Figure 3.4). Treatment decisions are made by artificial intelligence directly to patients or by caregivers depending on the treatment advice complexity [28].

Artificial intelligence was turned out to be a potential tool in image recognition regarding radiology, pathology, and recently ophthalmology [29]. It has been widely used and has been demonstrated to monitor fatigue [30]. For instance, the artificial neural network, data analysis of mouse interaction and keyboard patterns, and heart rate data examination have been investigated to identify the cause and features that help the recognition of fatigue. For the detection of COVID-19, Salman *et al.* [31] proposed a new model based on deep learning and X-ray imaging that will greatly improve the efficiency of radiologists and thus contribute to the control of the epidemic.

The authors of [32] presented a model-based logistic regression to group triple negative breast cancer on ultrasound images. The model provides a specificity of 82.91% and a sensitivity of 86.96% on the dataset used. Li *et al.* [33] used a probabilistic neural network to model diabetes diagnosis. The results obtained show that the model can economize doctors' time and improve diabetes diagnosis compared to the classical diagnosis process. In [34], the authors exploited the spectro-temporal changes of the electrocardiography signal to develop a recognition model using both generalized Morse wavelets and the short-time Fourier transform. The model reached a high average accuracy over the databases studied. In [35], a neural network is proposed to predict Parkinson's disease. The authors used a real-life dataset that included healthy and affected persons to conduct the experiments. The results showed an accuracy of 93.60% for the proposed technique and it outperforms traditional algorithms. Other important related artificial intelligence algorithms are shown in Table 3.1.

3.3 How to Develop Machine Learning Methods for Healthcare

Rapid advancement in machine learning methods allows for ameliorating healthcare decision support. However, the development of machine learning methods for healthcare involves special considerations to improve patient care. In this section, we present how to develop machine learning methods for healthcare. Importantly, we will focus on machine learning models for diagnosis, which involve the prediction of a label based on the input of health data. The paper [1] presented an outline of developing and implementing machine learning methods for healthcare and presented it in the form of five steps: selecting the appropriate problem and defining the prediction task, constructing data, developing models, evaluating model performance, and evaluating potential clinical impact.

3.3.1 Healthcare problem selection

For the development of machine learning models (or methods) for healthcare care, the first step is to select the appropriate problem and define the prediction

Table 3.1 Summary of some articles on the use of artificial intelligence techniques in healthcare

Paper	Purpose	Domain	Technique	Best performance	Limitations
Wu *et al.* [32]	This article uses machine learning methods to examine breast ultrasound images to provide a high level of diagnosis of triple negative breast cancer.	Radiology	Logistic regression	Specificity = 82.91% Sensitivity = 86.96%	Taking the age of the patients into account in the analysis did not improve the discrimination between triple-negative (TN) breast cancer and non-TN (NTN).
Li *et al.* [33]	This paper uses a backpropagation neural network and a probabilistic neural network to model diabetes diagnosis to provide doctors diagnosis and treatment with high precision.	Ophthalmology	Neural networks	Accuracy = 97.9%	Simply increasing the number of hidden layers of the backpropagation neural network will increase the computational complexity of the network and may not achieve the desired diagnosis.

Table 3.1   Continued

Paper	Purpose	Domain	Technique	Best performance	Limitations
Abdeldayem *et al.* [34]	This paper proposes an ECG-based biometric model for human identification. The extracted features have been enriched using generalized Morse wavelets and the short-time Fourier transform of the extracted ECG segments to obtain additional information on the ECG signal models.	Cardiology	Deep convolutional neural network	Accuracy = 97.85%	Some factors that are not taken into consideration can affect the ECG system, such as long-term factors including health and age status, while short-term factors include mental and emotional status.
Shrivastava *et al.* [35]	This paper demonstrates the potential of the Binary Bat algorithm to identify Parkinson's disease (PD) compared to classical methods.	Neurology	Neural network	Accuracy = 93.60%	This study does not allow one to differentiate PD from other Parkinsonian disorders such as rest tremor, akinesia, rigidity, and postural instability. Consequently, the doctor's intervention remains general and is not specific for each type of PD.

Table 3.1   Continued

Paper	Purpose	Domain	Technique	Best performance	Limitations
Patro *et al.* [36]	This paper proposes a biometric recognition method to select suitable features for features classification. The features are subjected to traditional classification algorithms to evaluate the performance of the proposed method.	Biometric human recognition	Least absolute shrinkage, artificial neural network, support vector machine, and *K*-nearest neighbor.	Accuracy = 99.1379%	The step of extracting clinical characteristics can have a significant impact on the performance of the biometric system.
Aggarwal *et al.* [37]	This paper implements the parameters of heart rate variation and investigates the support vector machine and the artificial neural network to develop an online automated real-time predictive model for diabetes.	Ophthalmology	Support vector machine and artificial neural network	Accuracy = 86.3%	The proposed online automated real-time prediction model for diabetes can only be used for the prognostic technique. Consequently, the system will not be able to replace the clinical decision and opinions of expert clinicians.

work. Effective machine learning models should play a significant effect in patient care by providing actionable knowledge and information. The models should contain a considerable number of parameters to better predictive performance.

3.3.2 Dataset construction

The second step is the construction of the dataset. However, there are always several challenges that need to be addressed, for example, lack of confident training data and contradictions in data formats. However, a new active research topic recently has been focused on optimizing the transfer of human learning to a machine learning model. As healthcare is appearing nowadays, researchers are concentrating on sorts of health data used for prediction [38]. Many search papers focus on clinical data [39], [40], [41], [42], but other papers, including sensors [43], [44], [45], and Omics data [46], [47], [48], [49], [50], [51], [52], [53], also use other data types. Data training and preparation are greatly significant and play a valuable role in the machine learning technique. It facilitates having a proper dataset before using any learning algorithm. Additionally, to apply machine learning algorithms in clinical data analysis, we should choose appropriate datasets, suitable model selection approaches, feature selection, and feature extraction. The types of healthcare data used for disease prediction and diagnosis are summarized in Figure 3.5.

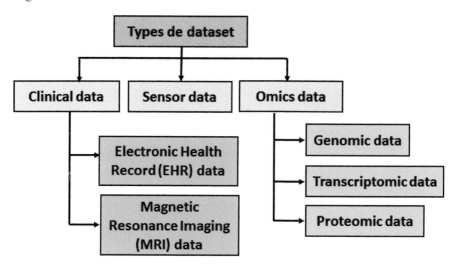

Figure 3.5   Types of healthcare data.

3.3.3 Model development

The development of machine learning models for healthcare and their successful deployment is not very evident and needs prudent estimation of the problems that could arise from different requirements [54]. Therefore, better machine learning models for healthcare must meet certain important criteria, such as transparency of diagnostic knowledge, the ability of the model to optimize the number of tests needed to make a reliable diagnosis, the ability to deal with missing data, the ability to provide good performance, and the ability to clarify and explain decisions made. As the health data is more sensible and any error can affect a person's life, the machine learning model developed must predict automatically the effect to save a person's life. For example, choices about the inputs and output data of a model are very indispensable in determining the practical value of the machine learning model in prediction and patient characteristics. Machine learning developers must ensure that the machine learning model only receives features of the patient's care and that the model uses those features.

3.3.4 Model performance evaluation

For evaluation of these models, many metrics are crucial, especially those that measure discrimination, using thresholds, and calibration, using probabilities. Particularly in critical use cases, in which it is necessary to make a clinical decision, there is certain hesitation in the deployment of machine learning models because the needed cost is high.

3.3.5 Clinical impact evaluation

The last step of the process deals with evaluating the potential clinical impact and validating the accuracy of the machine learning model in real-world scenarios. However, it should use interpretable machine learning models, which allow evaluating such a model before taking any decision. The lack of interpretability in machine learning models for healthcare can have life-threatening consequences in many cases. Interpretability is the fidelity and explanation of the model, and such a model should give more explanation of why it is making a prediction, making a decision, or giving a recommendation. To the machine learning model being exploited, it should be comprehensible to the health user and be applicable concerning the target use case.

3.4 Disease Prediction and Diagnosis Using Artificial Intelligence

3.4.1 *Disease prediction and diagnosis using machine learning*

Many studies in the literature focus on disease prediction and diagnosis using machine learning models. Authors in [2] highlighted the importance of machine learning models, as techniques of totally computer-guided health data analysis and interpretation. Dynamic clinical decisions develop a better patient–physician relationship and reduce healthcare costs [5], [55]. Furthermore, there are several health fields in which machine learning is highly being applied to early detection and diagnosis of diseases such as cancers [6], diabetic retinopathy [3], [9], [13], chronic diseases [11], cardiovascular diseases [8], vascular dementia [56], osteoporosis, heart disease [12], [53], Alzheimer's disease, and epilepsy. Machine learning applications know a lot of success in some areas that are under research such as radiology and emergency medicine [4], but cardiovascular systems have not received high success due to the lack of complete validation processes [7].

Authors in [11] discussed and proposed a novel multimodal disease risk prediction machine learning model for both unstructured and structured data. The results obtained in their work recorded a higher quality performance than those of the other existing methods. In addition, Zhao *et al.* [40] used machine learning to construct a predictive approach for the duration of the robot-assisted surgery (RAS) case. They concluded their study by a decrease in root mean square error to predict the lengths of the RAS cases by using different traditional machine learning models. The authors of [4] presented three learning techniques that capture hidden patterns in the data. The first technique deals with an unsupervised approach that can be chosen to discover key features, which drive differentiation between data samples. It includes approaches to K-means, spectral clustering, fuzzy clustering, and Gaussian mixture models. The second technique is the supervised method, which is a model building technique that determines characteristic sets of rules, which can be utilized to differentiate between different samples of classes. It includes the k-nearest neighbor, decision trees, support vector machines, and the Bayesian models' approach. The last technique deals with evolutionary methods and is referred to as genetic algorithms, which are based on evolutionary ideas of natural selection and the genetic processes of biological organisms.

Authors in [57] used explainable and interpretable machine learning models in healthcare to characterize the criteria to select the right explainable machine learning model for a given issue in healthcare. They presented many use cases across patient care to address the diverse nuances required to balance algorithmic optimization and explainability in problems such as the risk of readmission, admission to the emergency department, disease progression, and disease diagnosis. Furthermore, Kumar *et al.* [13] combined cloud computing and Internet of Things technologies in a framework composed of three phases focused on diabetes disease. It consists of data collection from the Internet of Things devices, storing the medical records securely in the cloud database and then predicting the disease with severity and diagnosing it. They proposed a new classification method, called a fuzzy rule-based neural classifier, for diagnosing diabetes disease and its severity. The proposed framework outperforms the existing models for disease prediction.

Table 3.2 summarizes the research papers with the best reported performance machine learning models for disease prediction and diagnosis in healthcare.

Table 3.2   Summary of existing important machine learning models for disease prediction and diagnosis in healthcare

Authors	Scope	Proposed model	Performance	Dataset
Ksia_zek *et al.* [58]	Detect hepatocellular carcinoma cancer (HCC) among adults.	A model used ten well-known machine learning algorithms to detect HCC using 165 patients.	Accuracy = 0.8849 F1-score = 0.8762	The HCC dataset was collected by the Coimbra University Hospital and Centre [59].
Abdar *et al.* [60]	Classify and automate the Wart disease treatment response.	A machine learning model based on the combination of improved adaptive particle swarm optimization algorithm and artificial immune recognition system.	Precision = 0.8908 Recall = 0.8943 F-measure = 0.8916 accuracy = 90%	The dataset contains 180 patients taken from immunotherapy and cryotherapy of Ghaem Hospital in Mashhad [61].

Table 3.2   Continued

Authors	Scope	Proposed model	Performance	Dataset
Acharya *et al.* [62]	Classify normal and coronary artery disease conditions using ECG signals.	A model based on the *k*-nearest neighbors classifier.	Accuracy = 98.17% Specificity = 99.34% Sensitivity = 94.57%	ECG-ID database [63].
Abdar *et al.* [64]	Detect coronary artery disease from Iranian patients.	The N2Genetic-NuSVM algorithm is based on a two-level hybrid genetic algorithm and NuSVM.	Accuracy = 93.08% *F*1-score = 91.51%	Z-Alizadeh Sani heart disease dataset [65].
Patro *et al.* [66]	Select appropriate features for classification and identification of ECG data.	Least absolute shrinkage and selection operator statistical learning model.	Accuracy = 99.1379%	ECG-ID database [63].
Ahmed *et al.* [67]	Predict heart disease.	Optimal real-time model based on random forest classifier.	Accuracy = 94.9%	Cleveland heart disease dataset [68].
Luo *et al.* [69]	Predict asthma disease in children.	A model to predict a child's asthma to provide real-time decision support.	Accuracy = 71.8% Sensitivity = 73.8% Specificity = 71.4%	The dataset was collected over two years on 210 children.

3.4.2 Artificial intelligence technology for clinical diagnosis

Manual diagnosis of diseases is often time-consuming and the chances of errors increase drastically. Artificial intelligence algorithms are a valuable means for assisting physicians in the diagnostic, prognostic, and pre- and post-treatment processes. A key task is to automatically determine the presence or absence of disease. This classification focuses on making clinical decisions on a type of disease or several classes of it, by detecting patterns corresponding to each class. There are various types of artificial intelligence methods that have been used for automatically diagnosing a diversity of different diseases, such as logistic regression, decision trees (DT), support vector machines (SVMs), neural networks (NN), artificial neural networks (ANN), and so forth. These artificial intelligence methods mostly use computer-based predictive analytics methods to classify, filter, organize, and try to find patterns in large datasets from diverse health sources. They are principally used as a means of detection or as a diagnostic aid allowing medical staff to make quick decisions. For example, convolutional neural networks make it possible to create image classification models for diagnosing lung and brain disorders. Statistical methods and random forest predictors allow the construction of risk models and survival estimators for heart disease to determine the patient's prognosis. Natural language processing allows information to be extracted from radiology reports.

Even though artificial intelligence technologies are invaluable and beneficial for patients and doctors and are applicable in different disease diagnoses, unfortunately, they also present many problems of a defective clinical diagnosis. The precise diagnosis of some diseases is based on several health data collected from many millions of patients who have experienced similar symptoms. To obtain better accuracy, the dataset used by the artificial intelligence method must have enough information about the patients. Therefore, if there is no information about a patient in a certain context, artificial intelligence can provide an inaccurate diagnosis. Consequently, the doctor will probably be able to give the wrong treatment to the patient if he does not have enough experience to identify the issue and it can cause the death of the patient. However, major problems need to be addressed for better practical use of artificial intelligence in disease diagnosis. Table 3.3 provides some explanations for how artificial intelligence technology can help in the clinical diagnosis corresponding to each disease.

Table 3.3   Artificial intelligence technology for clinical diagnosis.

Contribution	Technique	Disease	Clinical diagnosis	Limitation
Wang *et al.* [70]	Support vector machines	Breast cancer	- Increase the accuracy of breast cancer. - Reduce variance of diagnostic accuracy of breast cancer.	Structures and configuration parameters of the SVM technique can affect the performance of diagnosis.
Reddy *et al.* [71]	Genetic algorithm and fuzzy logic	Heart	- Reduce the risk of patient death and the cost of treatment. - Helps doctors diagnose patients heart disease at early stages.	The used environment is static and health data is not streaming.
Jeyaranjani *et al.* [72]	Artificial neural networks	Coronary heart	- It is efficient for the doctor to predict the stages of coronary heart disease. - Diagnoses misclassified coronary heart patient records. - Shows the angiographic disease status of the patient. - Reduces the risk of late diagnosis, which saves human lives.	More execution time is needed to generate better results.
Thammastitkul *et al.* [73]	Naive Bayesian	Diabetic retinopathy	- Early detection of microaneurysms in diabetic retinopathy to reduce the occurrence of blindness.	Not tested on a large dataset.

Table 3.3   Continued

Contribution	Technique	Disease	Clinical diagnosis	Limitation
Castellazzi *et al.* [74]	Machine learning with magnetic resonance imaging	Alzheimer's disease (AD) and vascular dementia (VD)	- Identify particular regions of the brain such as the hippocampus, anterior cingulum, and thalamus with clearly defined types of dementia. - Provide early prediction of disease type (AD or DV) in patients with clinical symptoms of mixed dementia. - Support and assist clinicians to improve diagnostic and prognostic accuracy as well as therapy and patient management.	Low performance with a unimodal feature dataset.
Yamamoto *et al.* [75]	Deep learning	Osteoporosis	- High-accuracy diagnosis of osteoporosis from hip radiographs. - Improve diagnosis performance by using clinical covariates from patient records.	Computationally complex.

Table 3.3   Continued

Contribution	Technique	Disease	Clinical diagnosis	Limitation
Yu *et al.* [76]	Random forest and long short-term memory	Stroke	- Provide real-time detection of stroke using real-time biosignals to allow initial changes in the patient's physiological status for stroke therapeutic purposes. - Warn medical staff and doctors to preemptively respond to the healthcare needs of older people.	The model does not take into account the cases of driving, sleeping, or walking during everyday life to collect real-time biosignals.
Jin *et al.* [77]	Facial expression recognition	Parkinson	- Help patients achieve more full treatment. - Help physicians understand the real-time dynamics of Parkinson's disease.	The used dataset was not sufficient.
Abiyev *et al.* [78]	Convolutional neural networks	Epilepsy	- Classify EEG signals into a seizure and normal classes to study brain activity features.	Computationally complex.

3.5 Challenges of using Artificial Intelligence Algorithms in Healthcare

Machine learning provides many approaches and techniques that help in the diagnostic and prognostic challenges faced in healthcare. Even though machine learning models for healthcare have many advantages in disease prediction and diagnosis, unfortunately, they are not general and standard-perfect models. The following parameters summarize the problems encountered when we want to develop machine learning models for healthcare disease prediction and diagnosis [79], [80]:

- There are significant challenges that occur with fulfillment to data, which should influence any decision related to the implementation of machine learning models in healthcare. Most health data are unstructured data and come in the form of various types, such as images of discharge summaries, reports, and audio or video recordings. Therefore, it is very hard to categorize and quantify such data [51].
- The quantity of data generated by patients is exponentially increasing and is becoming more difficult to manage in the timeframes necessary to be clinically helpful.
- Different machine learning models are suitable for a certain problem for a specific healthcare dataset and cannot work well on some other datasets. Therefore, selecting an appropriate and good machine learning solution for a particular dataset is a very big challenge in the healthcare field.
- Machine learning models for healthcare need huge, high-quality, and big datasets to be trained. However, these datasets need more time to be collected. Data availability is considered among the biggest challenges for the development and evaluation of machine learning algorithms for healthcare.
- Machine learning models for healthcare care need time, many resources, and computer equipment to generate high-confidence results. It is very difficult to prove that the prediction made by these models works correctly for all possible scenarios in a real-time application. All depend on the information technology resources and the equipment used.
- The quality of the health dataset is not always good so that we have to spend a lot of time cleaning the data and working on new techniques to use the health data efficiently. This sometimes takes more time to make decisions that are not good in the area of health, which is very sensitive to delay.
- It is very difficult to provide proper interpretation and the right analysis of the generated results by machine learning models.

Conclusion

The development of artificial intelligence in healthcare care is still a burgeoning field. In recent years, related research emerged for the benefit of different diseases. As it knows successful results and benefits, it has some disadvantages and limitations. However, efforts are still being made to develop artificial intelligence techniques to save more lives. Moreover, in this chapter, we have focused on the utilization of some artificial intelligence in healthcare applications to provide better and remote solutions, especially in the case of disease identification and diagnosis. Furthermore, we have concentrated on which type of artificial intelligence technique we can use on what type of diseases a person currently facing in the healthcare field. After that, we have presented the steps to develop machine learning models for healthcare and a brief overview of the challenges experienced when deploying these models in the healthcare industry. Finally, we have presented some challenges caused to develop models for healthcare disease diagnosis.

References

[1]   Chen, P. H. C., Liu, Y., & Peng, L. (2019). How to develop machine learning models for healthcare. *Nature materials*, *18*(5), 410-414.

[2]   Beam, A. L., & Kohane, I. S. (2018). Big data and machine learning in health care. *Jama*, *319*(13), 1317-1318.

[3]   Gulshan, V., Peng, L., Coram, M., Stumpe, M. C., Wu, D., Narayanaswamy, A., ... & Webster, D. R. (2016). Development and validation of a deep learning algorithm for detection of diabetic retinopathy in retinal fundus photographs. *Jama*, *316*(22), 2402-2410.

[4]   Mansi, T., Mihalef, V., Sharma, P., Georgescu, B., Zheng, X., Rapaka, S., ... & Comaniciu, D. (2012, May). Data-driven computational models of heart anatomy, mechanics and hemodynamics: An integrated framework. In *2012 9th IEEE International Symposium on Biomedical Imaging (ISBI)* (pp. 1434-1434). IEEE.

[5]   Ngiam, K. Y., & Khor, W. (2019). Big data and machine learning algorithms for health-care delivery. *The Lancet Oncology*, *20*(5), e262-e273.

[6]   Belle, A., Kon, M. A., & Najarian, K. (2013). Biomedical informatics for computer-aided decision support systems: a survey. *The Scientific World Journal*, *2013*.

[7]   Caceres, C. A., & Rikli, A. E. (1961). The digital computer as an aid in the diagnosis of cardiovascular disease. *Transactions of the New York Academy of Sciences*, *23*(3 Series II), 240-245.

[8] Martis, R. J., Chakraborty, C., & Ray, A. K. (2014). Wavelet-based machine learning techniques for ECG signal analysis. In *Machine learning in healthcare informatics* (pp. 25-45). Springer, Berlin, Heidelberg.

[9] Zhu, K. Y., Liu, W. D., & Xiao, Y. (2014). Application of fuzzy logic control for regulation of glucose level of diabetic patient. In *Machine Learning in Healthcare Informatics* (pp. 47-64). Springer, Berlin, Heidelberg.

[10] Hamet, P., & Tremblay, J. (2017). Artificial intelligence in medicine. *Metabolism*, *69*, S36-S40.

[11] Chen, M., Hao, Y., Hwang, K., Wang, L., & Wang, L. (2017). Disease prediction by machine learning over big data from healthcare communities. *Ieee Access*, *5*, 8869-8879.

[12] Wu, C. S. M., Badshah, M., & Bhagwat, V. (2019, July). Heart disease prediction using data mining techniques. In *Proceedings of the 2019 2nd international conference on data science and information technology* (pp. 7-11).

[13] Kumar, P. M., Lokesh, S., Varatharajan, R., Babu, G. C., & Parthasarathy, P. (2018). Cloud and IoT based disease prediction and diagnosis system for healthcare using Fuzzy neural classifier. *Future Generation Computer Systems*, *86*, 527-534.

[14] Garbuio, M., & Lin, N. (2019). Artificial intelligence as a growth engine for health care startups: Emerging business models. *California Management Review*, *61*(2), 59-83.

[15] Rong, G., Mendez, A., Assi, E. B., Zhao, B., & Sawan, M. (2020). Artificial intelligence in healthcare: review and prediction case studies. *Engineering*, *6*(3), 291-301.

[16] Panch, T., Szolovits, P., & Atun, R. (2018). Artificial intelligence, machine learning and health systems. *Journal of global health*, *8*(2).

[17] Shen, D., Wee, C. Y., Zhang, D., Zhou, L., & Yap, P. T. (2014). Machine learning techniques for AD/MCI diagnosis and prognosis. In *Machine learning in healthcare informatics* (pp. 147-179). Springer, Berlin, Heidelberg.

[18] El Kafhali, S., & Salah, K. (2017). Efficient and dynamic scaling of fog nodes for IoT devices. *The Journal of Supercomputing*, *73*(12), 5261-5284.

[19] El Kafhali, S., Salah, K., & Ben Alla, S. (2018, November). Performance evaluation of IoT-fog-cloud deployment for healthcare services. In *2018 4th international conference on cloud computing technologies and applications (Cloudtech)* (pp. 1-6). IEEE.

[20] El Kafhali, S., & Salah, K. (2019). Performance modelling and analysis of Internet of Things enabled healthcare monitoring systems. *IET Networks*, 8(1), 48-58.

[21] Reddy, S., Allan, S., Coghlan, S., & Cooper, P. (2020). A governance model for the application of AI in health care. *Journal of the American Medical Informatics Association*, 27(3), 491-497.

[22] Shameer, K., Johnson, K. W., Glicksberg, B. S., Dudley, J. T., & Sengupta, P. P. (2018). Machine learning in cardiovascular medicine: are we there yet?. *Heart*, 104(14), 1156-1164.

[23] Randhawa, G. S., Soltysiak, M. P., El Roz, H., de Souza, C. P., Hill, K. A., & Kari, L. (2020). Machine learning using intrinsic genomic signatures for rapid classification of novel pathogens: COVID-19 case study. *Plos one*, 15(4), e0232391.

[24] Ge, Y., Tian, T., Huang, S., Wan, F., Li, J., Li, S., ... & Zeng, J. (2020). A data-driven drug repositioning framework discovered a potential therapeutic agent targeting COVID-19. *BioRxiv*.

[25] Senior, A. W., Evans, R., Jumper, J., Kirkpatrick, J., Sifre, L., Green, T., ... & Hassabis, D. (2020). Improved protein structure prediction using potentials from deep learning. *Nature*, 577(7792), 706-710.

[26] Alimadadi, A., Aryal, S., Manandhar, I., Munroe, P. B., Joe, B., & Cheng, X. (2020). Artificial intelligence and machine learning to fight COVID-19. *Physiological genomics*, 52(4), 200-202.

[27] Khan, A. I., Shah, J. L., & Bhat, M. M. (2020). CoroNet: A deep neural network for detection and diagnosis of COVID-19 from chest x-ray images. *Computer Methods and Programs in Biomedicine*, 196, 105581.

[28] El Kafhali, S., & Lazaar, M. (2021). Artificial Intelligence for Healthcare: Roles, Challenges, and Applications. In *Intelligent Systems in Big Data, Semantic Web and Machine Learning* (pp. 141-156). Springer, Cham.

[29] Kuht, H., Wang, S., Nishad, G., George, S., Maconachie, G., Sheth, V., ... & Thomas, M. G. (2020). Using Artificial Intelligence (AI) to Classify Retinal Developmental Disorders. *Investigative Ophthalmology & Visual Science*, *61*(7), 4030-4030.

[30] Parekh, V., Shah, D., & Shah, M. (2020). Fatigue detection using artificial intelligence framework. *Augmented Human Research*, *5*(1), 1-17.

[31] Salman, F. M., Abu-Naser, S. S., Alajrami, E., Abu-Nasser, B. S., & Alashqar, B. A. (2020). Covid-19 detection using artificial intelligence, 4, 18-25.

[32] Wu, T., Sultan, L. R., Tian, J., Cary, T. W., & Sehgal, C. M. (2019). Machine learning for diagnostic ultrasound of triple-negative breast cancer. *Breast cancer research and treatment*, *173*(2), 365-373.

[33] Li, X. (2021). Artificial intelligence neural network based on intelligent diagnosis. *Journal of Ambient Intelligence and Humanized Computing*, *12*, 923-931.

[34] Abdeldayem, S. S., & Bourlai, T. (2018, December). ECG-based human authentication using high-level spectro-temporal signal features. In *2018 IEEE international conference on big data (big data)* (pp. 4984-4993). IEEE.

[35] Shrivastava, P., Shukla, A., Vepakomma, P., Bhansali, N., & Verma, K. (2017). A survey of nature-inspired algorithms for feature selection to identify Parkinson's disease. *Computer methods and programs in biomedicine*, *139*, 171-179.

[36] Patro, K. K., Reddi, S. P. R., Khalelulla, S. E., Kumar, P. R., & Shankar, K. (2020). ECG data optimization for biometric human recognition using statistical distributed machine learning algorithm. *The Journal of Supercomputing*, *76*(2), 858-875.

[37] Aggarwal, Y., Das, J., Mazumder, P. M., Kumar, R., & Sinha, R. K. (2020). Heart rate variability features from nonlinear cardiac dynamics in identification of diabetes using artificial neural network and support vector machine. *Biocybernetics and Biomedical Engineering*, *40*(3), 1002-1009.

[38] Dhillon, A., & Singh, A. (2019). Machine learning in healthcare data analysis: a survey. *Journal of Biology and Today's World*, *8*(6), 1-10.

[39] Zheng, T., Xie, W., Xu, L., He, X., Zhang, Y., You, M., ... & Chen, Y. (2017). A machine learning-based framework to identify type 2 diabetes through electronic health records. *International journal of medical informatics*, *97*, 120-127.

[40] Corey, K. M., Kashyap, S., Lorenzi, E., Lagoo-Deenadayalan, S. A., Heller, K., Whalen, K., ... & Sendak, M. (2018). Development and validation of machine learning models to identify high-risk surgical patients using automatically curated electronic health record data (Pythia): a retrospective, single-site study. *PLoS medicine*, *15*(11), e1002701.

[41] Rahimian, F., Salimi-Khorshidi, G., Payberah, A. H., Tran, J., Ayala Solares, R., Raimondi, F., ... & Rahimi, K. (2018). Predicting the risk of emergency admission with machine learning: Development and validation using linked electronic health records. *PLoS medicine*, *15*(11), e1002695.

[42] Panahiazar, M., Taslimitehrani, V., Pereira, N., & Pathak, J. (2015). Using EHRs and machine learning for heart failure survival analysis. *Studies in health technology and informatics*, *216*, 40.

[43] Kaur, P., & Malhotra, S. (2019). Improved SLReduct Framework for Stress Detection Using Mobile Phone-Sensing Mechanism in Wireless Sensor Network. In *Progress in Advanced Computing and Intelligent Engineering* (pp. 499-507). Springer, Singapore.

[44] Kanjo, E., Younis, E. M., & Ang, C. S. (2019). Deep learning analysis of mobile physiological, environmental and location sensor data for emotion detection. *Information Fusion*, *49*, 46-56.

[45] Hassan, M. M., Huda, S., Uddin, M. Z., Almogren, A., & Alrubaian, M. (2018). Human activity recognition from body sensor data using deep learning. *Journal of medical systems*, *42*(6), 1-8.

[46] Njage, P. M. K., Leekitcharoenphon, P., & Hald, T. (2019). Improving hazard characterization in microbial risk assessment using next generation sequencing data and machine learning: Predicting clinical outcomes in shigatoxigenic Escherichia coli. *International journal of food microbiology*, *292*, 72-82.

[47] Gurovich, Y., Hanani, Y., Bar, O., Nadav, G., Fleischer, N., Gelbman, D., ... & Gripp, K. W. (2019). Identifying facial phenotypes of genetic disorders using deep learning. *Nature medicine*, *25*(1), 60-64.

[48] Bobak, C. A., Titus, A. J., & Hill, J. E. (2019). Comparison of common machine learning models for classification of tuberculosis using transcriptional biomarkers from integrated datasets. *Applied Soft Computing, 74*, 264-273.

[49] Alakwaa, F. M., Chaudhary, K., & Garmire, L. X. (2018). Deep learning accurately predicts estrogen receptor status in breast cancer metabolomics data. *Journal of proteome research, 17*(1), 337-347.

[50] Mamoshina, P., Volosnikova, M., Ozerov, I. V., Putin, E., Skibina, E., Cortese, F., & Zhavoronkov, A. (2018). Machine learning on human muscle transcriptomic data for biomarker discovery and tissue-specific drug target identification. *Frontiers in genetics, 9*, 242.

[51] Liang, C. A., Chen, L., Wahed, A., & Nguyen, A. N. (2019). Proteomics analysis of FLT3-ITD mutation in acute myeloid leukemia using deep learning neural network. *Annals of Clinical & Laboratory Science, 49*(1), 119-126.

[52] Panda, B., & Majhi, B. (2021). A novel improved prediction of protein structural class using deep recurrent neural network. *Evolutionary Intelligence, 14*(2), 253-260.

[53] Srinivasan, S., Leshchyk, A., Johnson, N. T., & Korkin, D. (2020). A hybrid deep clustering approach for robust cell type profiling using single-cell RNA-seq data. *RNA, 26*(10), 1303-1319.

[54] Jackson, S., Yaqub, M., & Li, C. X. (2019). The agile deployment of machine learning models in healthcare. *Frontiers in big Data, 1*, 7.

[55] Bhardwaj, R., Nambiar, A. R., & Dutta, D. (2017, July). A study of machine learning in healthcare. In *2017 IEEE 41st Annual Computer Software and Applications Conference (COMPSAC)* (Vol. 2, pp. 236-241). IEEE.

[56] Iadecola, C. (2013). The pathobiology of vascular dementia. *Neuron, 80*(4), 844-866.

[57] Ahmad, M. A., Eckert, C., & Teredesai, A. (2018, August). Interpretable machine learning in healthcare. In *Proceedings of the 2018 ACM international conference on bioinformatics, computational biology, and health informatics* (pp. 559-560).

[58] Książek, W., Abdar, M., Acharya, U. R., & Pławiak, P. (2019). A novel machine learning approach for early detection of hepatocellular carcinoma patients. *Cognitive Systems Research, 54*, 116-127.

[59] Santos, M. S., Abreu, P. H., García-Laencina, P. J., Simão, A., & Carvalho, A. (2015). A new cluster-based oversampling method for improving survival prediction of hepatocellular carcinoma patients. *Journal of biomedical informatics*, *58*, 49-59.

[60] Abdar, M., Wijayaningrum, V. N., Hussain, S., Alizadehsani, R., Plawiak, P., Acharya, U. R., & Makarenkov, V. (2019). IAPSO-AIRS: A novel improved machine learning-based system for wart disease treatment. *Journal of medical systems*, *43*(7), 1-23.

[61] Khozeimeh, F., Alizadehsani, R., Roshanzamir, M., Khosravi, A., Layegh, P., & Nahavandi, S. (2017). An expert system for selecting wart treatment method. *Computers in biology and medicine*, *81*, 167-175.

[62] Acharya, U. R., Sudarshan, V. K., Koh, J. E., Martis, R. J., Tan, J. H., Oh, S. L., ... & San Tan, R. (2017). Application of higher-order spectra for the characterization of coronary artery disease using electrocardiogram signals. *Biomedical Signal Processing and Control*, *31*, 31-43.

[63] Goldberger, A. L., Amaral, L. A., Glass, L., Hausdorff, J. M., Ivanov, P. C., Mark, R. G., ... & Stanley, H. E. (2000). PhysioBank, PhysioToolkit, and PhysioNet: components of a new research resource for complex physiologic signals. *circulation*, *101*(23), e215-e220.

[64] Abdar, M., Książek, W., Acharya, U. R., Tan, R. S., Makarenkov, V., & Pławiak, P. (2019). A new machine learning technique for an accurate diagnosis of coronary artery disease. *Computer methods and programs in biomedicine*, *179*, 104992.

[65] Frank, A., & Asuncion, A. (2018). UCI Machine Learning Repository: Center for Machine Learning and Intelligent Systems. https: //archive. ics.uci.edu/ml/datasets/Z-Alizadeh+Sani.

[66] Patro, K. K., Reddi, S. P. R., Khalelulla, S. E., Kumar, P. R., & Shankar, K. (2020). ECG data optimization for biometric human recognition using statistical distributed machine learning algorithm. *The Journal of Supercomputing*, *76*(2), 858-875.

[67] Ahmed, H., Younis, E. M., Hendawi, A., & Ali, A. A. (2020). Heart disease identification from patients' social posts, machine learning solution on Spark. *Future Generation Computer Systems*, *111*, 714-722.

[68] S. H. Cleveland, The va long beach, (2019). heart disease data set, https://archive.ics.uci.edu/ml/datasets/heart+Disease.

[69] Luo, G., Stone, B. L., Fassl, B., Maloney, C. G., Gesteland, P. H., Yerram, S. R., & Nkoy, F. L. (2015). Predicting asthma control deterioration in children. *BMC medical informatics and decision making, 15*(1), 1-8.

[70] Wang, H., Zheng, B., Yoon, S. W., & Ko, H. S. (2018). A support vector machine-based ensemble algorithm for breast cancer diagnosis. *European Journal of Operational Research, 267*(2), 687-699.

[71] Reddy, G. T., Reddy, M. P. K., Lakshmanna, K., Rajput, D. S., Kaluri, R., & Srivastava, G. (2020). Hybrid genetic algorithm and a fuzzy logic classifier for heart disease diagnosis. *Evolutionary Intelligence, 13*(2), 185-196.

[72] Jeyaranjani, J., Rajkumar, T. D., & Kumar, T. A. (2021). Coronary heart disease diagnosis using the efficient ANN model. *Materials Today: Proceedings.* https://doi.org/10.1016/j.matpr.2021.01.257

[73] Thammastitkul, A., Uyyanonvara, B., & Barman, S. A. (2020). Improving microaneurysm detection from non-dilated diabetic retinopathy retinal images using feature optimisation. *International Journal of Computer Aided Engineering and Technology, 12*(3), 355-369. 12(3), 355-369.

[74] Castellazzi, G., Cuzzoni, M. G., Cotta Ramusino, M., Martinelli, D., Denaro, F., Ricciardi, A., ... & Gandini Wheeler-Kingshott, C. A. (2020). A machine learning approach for the differential diagnosis of Alzheimer and Vascular Dementia Fed by MRI selected features. *Frontiers in neuroinformatics, 14*, 25.

[75] Yamamoto, N., Sukegawa, S., Kitamura, A., Goto, R., Noda, T., Nakano, K., ... & Ozaki, T. (2020). Deep learning for osteoporosis classification using hip radiographs and patient clinical covariates. *Biomolecules, 10*(11), 1534.

[76] Yu, J., Park, S., Kwon, S. H., Ho, C. M. B., Pyo, C. S., & Lee, H. (2020). AI-Based Stroke Disease Prediction System Using Real-Time Electromyography Signals. *Applied Sciences, 10*(19), 6791.

[77] Jin, B., Qu, Y., Zhang, L., & Gao, Z. (2020). Diagnosing Parkinson disease through facial expression recognition: video analysis. *Journal of Medical Internet Research, 22*(7), e18697.

[78] Abiyev, R., Arslan, M., Bush Idoko, J., Sekeroglu, B., & Ilhan, A. (2020). Identification of epileptic EEG signals using convolutional neural networks. *Applied Sciences, 10*(12), 4089.

[79] Alizadehsani, R., Abdar, M., Roshanzamir, M., Khosravi, A., Kebria, P. M., Khozeimeh, F., ... & Acharya, U. R. (2019). Machine learning-based coronary artery disease diagnosis: A comprehensive review. *Computers in biology and medicine*, *111*, 103346.

[80] Rowe, M. (2019). An introduction to machine learning for clinicians. *Academic Medicine*, *94*(10), 1433-1436.

4

Segmentation of MRI Images of Gliomas using Convolutional Neural Networks

Anupama M. Nayak[1], R. Rachana[1], M. M. Vishakh[2],
S. C. Prasanna Kumar[2], Praveen Kumar Gupta[1],
Sumathra Manokaran[1], and A. H. Manjunatha Reddy[#1]

[1]Department of Biotechnology, R V College of Engineering, India
[2]Department of Electronics and Instrumentation Technology, R V College of Engineering, India;
[#]Corresponding author: ahmanjunatha@rvce.edu.in, Phone: 9844573697

Abstract

The need for accurate and intuitive diagnostic tools in the field of biomedical science has incited many researchers to focus on integrating the various models offered by deep learning to suit therapeutic utility. Neural networks, in particular, have been exhaustively worked upon to aid in image segmentation and classification operations. This project aims to address the need for automatic segmentation of MRI images of gliomas in cancer patients. The DenseNet architectural variant of convolutional neural networks has been utilized to build a highly accurate 3D segmentation tool. Our program also functions to perform preliminary classification of the tumors into high-grade tumors and low-grade tumors to afford the medical community an initial insight into the severity of the gliomas. We have utilized the open-source Python library function to build our architecture. The training and testing of this architecture were performed on the benchmark MICCAI BRaTS dataset. Our model showed high segmentation precision where we obtained accuracy close to 100% (~99.94% with enhancements). The same values were obtained for sensitivity and positive predictive values (PPVs). This level of segmentation accuracy can be considered ideal for practical medical use.

4.1 Introduction

Gliomas are a category of brain and spinal cord tumors that arise from anomalous division of glial cells [1]. They are known to infiltrate the surrounding tissues which make surgical removal and treatment of these tumors very hard, resulting in a high casualty rate for those afflicted by these tumors. These tumors can be malignant or benign. WHO (World Health Organization) has a segregation system that classifies these tumors with a I–IV grading system with IV representing the most aggressive form [2]. Gliomas are classified depending on the specific type of brain cell that is affected. By introducing the computerized tomography (CT) and magnetic resonance imaging (MRI) techniques, the brain tumor incidence rates have been relatively stable.

Despite numerous developments in glioma research, the patient diagnoses are very poor. More aggressive form of the disease is being considered as high-grade, which require immediate treatment. They have median survival rate of less than two years. The low-grade variants, which are slow-growing, *viz.*, low-grade astrocytomas or oligodendrogliomas, have relatively higher life expectancy, which allows the delay of overly aggressive treatment methods as they can cause serious tissue damage. Less than 3% of glioblastoma patients are still alive at 5 years after diagnosis [3]. Early detection of the brain tumor is crucial and has been the crux for a lot of research.

Neuroimaging is one of the important tools for diagnosis, treatment planning, and post-therapy assessment of any brain tumors. Radiology has played an integral role in this process and a lot of progress has been made in this field to involve accurate diagnosis, non-invasively and assessing the therapeutic response [4]. The brain tumor segmentation helps in separating the tumor from the normal brain cells, which will provide the information for diagnosis and planning can be emphasized for treatment.

The magnetic resonance imaging (MRI) techniques have been conventionally employed in modern neuroimaging for its superior characterization of structure and its ability to capture various cellular and metabolic functions [5], [6], [7]. The standardized neuroimaging protocol includes, three-dimensional (3D) T1, axial fluid-attenuated inversion recovery (FLAIR), axial diffusion-weighted imaging (DWI), axial gadolinium contrast-enhanced T2 and 3D gadolinium contrast-enhanced T1. MRI offers a higher contrast based image which enables better visualization and segmentation which is why these images have been employed in most segmentation algorithms. Accurate segmentation is vital to glioma study and diagnosis. Manual segmentation is a labor inductive method that is prone to misinterpretation and variability. Deep learning systems have been exploited by physicians to aid in offering second opinions and flagging regions of interest (ROIs) in images [8]. Automatic segmentation using 3D CNN (convolutional neural networks) has shown high efficiency in capturing multi-scale contextual information from the images [9]. The process involves the hierarchical segmentation of the entire tumor structure, core region, and enhanced tumor structures.

Convolutional neural networks (CNNs) are deep learning algorithms designed to process the data of natural spatial invariance. In CNN, the receptive field of a node is the region of the input signal that is involved in the multilayer convolution up to that node in a forward pass [10], [11]. The composition of a typical segmentation architecture is as follows:

1. Downsampling path, which is responsible for extracting coarse semantic features
2. Upsampling path trained to recover the input image resolution at the output of the model and optionally
3. Post-processing module for the refinement of the model predictions.

Densely connected convolutional networks or DenseNets have produced great outputs when utilized for image classification tasks [12]. DenseNets are based on directly connecting each and every layer to the other layer in feed-forward fashion to make the network more accurate and increase the ease of training the network. They are built using pooling operations and dense blocks where every block is one of the iterative concatenations of previous feature maps. This architecture is an extension of residual networks (ResNets) and designed for additional uses such as the following:

1. Parameter efficiency: DenseNets use the parameters more efficiently
2. Deep supervision: DenseNets perform deep supervision using the short paths in the architecture
3. Feature reuse: The current layer under study utilizes the previous layer information with computed feature maps. These attributes make them a good fit for the semantic segmentation.

The output of the "lth" layer, x_1 is computed by applying nonlinear transformation, "H" to the output of the previous layer, and x_{l-1} in a standard CNN. H is defined as convolution followed by rectified non-linear function (ReLU) and a dropout. The output is calculated as $x = H_1(x_1 - 1)$.

DenseNets integrate a pattern with more connectivity with all feature outputs in a feed-forward fashion. In this care, the output of the "*l*th" layer is given by $x_1 = H_1 ([x_{1-1}, x_{1-2}, ..., x_0])$, where the concatenation operation is represented. Here, reuse of the features will occur and all layers of the architecture receive supervision signals. For each layer, the output dimension has k feature maps where k is the growth rate parameter and k is typically set to a small value. Hence, the number of feature maps in DenseNets grows linearly with the depth.

This makes the DenseNet architecture suitable for efficient and accurate segmentation of the images while preserving the essential edge details. As per the segmentation performed, the neural network is to segregate the images into low grade gliomas. This gives a measure of the severity of the gliomas in accordance with the WHO classification of gliomas.

4.2 Aim

This study aims to tackle the automation of the segmentation of large amounts of MRI data produced while imaging low grade and high grade gliomas, using DenseNet architecture. We endeavor to aid in gliomal treatment at the diagnostic level, addressing the crucial task of accurate and efficient segmentation and gradation of gliomal tumors. More precise segmentation of brain tumors that resulted from MRI images reveals crucial and challenging tasks in diagnosis and planning for tumor treatment. Along the boundaries of tumors, it is still a challenging task because of the irregularities. Advancements in this area of research are the use of convolutional neural networks (CNNs). We take on the essential task of improving classification accuracy by extracting more efficient features while using minimal training data. There is still a lot of room for additional research to implement various machine learning techniques and modify them to specifically address the requirements of tumor diagnosis. These techniques should also satisfy the criteria of generic application in order for them to have utility in the medical field and satisfy general requirements of patient–doctor use.

4.3 Objectives

- Building a DenseNet architecture for MRI image segmentation
- Neural network training for segmentation of brain tumor
- Testing the trained neural network to obtain segmented images
- Classification of tumors as high grade (HGG) or low grade gliomas (LGG).

4.4 Methodology

- The MRI images obtained using the BRaTS dataset are pre-processed and are used for training and testing of the neural network.
- The ReLU activation function was utilized to linearize the input files.
- Convolutions functions were then employed on the linearized image files to obtain feature maps.
- These feature maps were inputted into the DenseNet.
- In the DenseNet, they passed through the dense block first where a separate set of activation and convolution functions was applied.
- Connecting each dense block was a transition layer which had its own set of linearizing and convolution functions.
- This process was repeated to enable a two-stage network.
- The output from these was subjected to Softmax functions to obtain the final regions of interest (ROIs).

This gave a trained DenseNet architecture to segment glioma images. Figure 4.1 shows the steps mentioned above for training.

- The training set of images from the BRaTS dataset was then allowed to undergo the same pre-processing steps.
- They were fed to the trained neural network to obtain accuracy scores.
- The obtained ground truth images were processed using an image editing program.
- The post-processing of the images involved annotating RGB values to the gliomal tissue.
- The images were then classified into LGG and HGG.

Resulting image frames were then saved and visualized using an image visualization software. Figure 4.2 shows the steps mentioned above for testing.

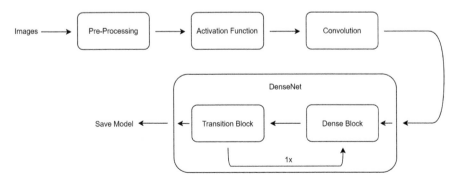

Figure 4.1 Training block diagram.

Figure 4.2   Testing block diagram.

4.5 Implementation

The entire project was implemented using Python programming language. Many open source libraries were used, with the most important one being TensorFlow.

The project was executed in a series of three separate steps which were later integrated. The four phases of creating a dense neural network to segment the MRI images of the gliomas composed of the following.

4.5.1 Initial image processing

The dataset of MRI gliomal images for the training and testing of the data was obtained from the MICCAI BRaTS 2018 dataset. The individual images from the dataset were taken in a patch size of 40 × 40 × 40 for the pre-processing. Histogram equalization was conducted in order to further extract local areas with better contrast features to produce higher edge resolution. Activation functions were then utilized to linearize the input values before feeding them to the network. The choice of activation functions is integral to determining the training efficiency of a model and its ability and speed of convergence and implemented ReLU (rectified linear unit) activation function. They are most widely used in CNNs where they linearize the variables through the mathematical function, $y = \max(0, x)$. Finally, a convolution step was performed where the output was found to be the most efficient and accurate at 18 layers. The 3D convolution was done to obtain an impulse response among output and the input signals.

DenseNet architecture was implemented for training and testing the CNN to segment the MRI images using the BRaTS dataset and obtain the regions of interest.

4.5.2 Training phase

To train the dense neural network, 20 datasets were used from the BRaTS-2018 dataset. Each dataset had been iterated 155 times in order to obtain the maximum accuracy. Apart from this, each dataset was allowed for five epochs. This architecture was implemented in a series of two blocks which were run as a two-stage network. The parameters are reduced despite the concatenation of several residuals by using bottlenecks. The images are first segmented with a dense block and later followed by a transition block. The dimensions of the feature maps does not vary within a dense block, while the number of filters changes amongst them, which are known as transition layers, and they perform the downsampling.

The dense block is composed of three parts containing 18 layers with 9 filters. Each layer in this block receives the entire output feature maps of the previous layers. Growth rate of dense blocks was set to 12. This parameter regulates the amount of information given to the network in each layer. A layer of a dense block constituted of the following:

- Batch normalization
- ReLU activation
- $3 \times 3 \times 3$ convolution

This entire sequence is carried out six times and the output of each individual process is then concatenated to obtain the mean segmentation of the image.

Following the dense block, the operation is then carried out in the transition layer. The transition layer generates $\lfloor \theta_m \rfloor$, $0 < \theta < 1$ output feature-map, where θ is the compression factor and m is the number of output feature maps. It allows compression of the model by making it more compact. This process plays a role in removing feature redundancy by sharing all the learned features. This again was divided into the sub-processes of the following:

- Batch normalization
- ReLU activation
- $1 \times 1 \times 1$ convolution

The first layer in this case was composed of 92 filters, while the second layer was made up of 168 filters.

The outputs from these dense and transition blocks were then inputted into a Softmax activation function for the segmentation and extraction of the region of interest (ROI).

4.5.3 Testing phase

The training phase used 40 images, containing both HGG and LGG in random order. This phase incorporated the same process of using the two-stage networks consisting of layered dense blocks interspersed with transition layers to contribute to image compression.

The dense blocks in this phase consisted of the 18 layers each equipped with 9 filters which segment the images through the following:

- Batch normalization
- ReLU activation
- $3 \times 3 \times 3$ convolution

The transition layers interspersed between the dense blocks had three further phases:

- Batch normalization
- ReLU activation
- $1 \times 1 \times 1$ convolution

This code was further augmented to obtain scores that would give a measure of the accuracy and the sensitivity and the specificity of the code. For the accuracy scores, we integrated the Dice function.

4.5.4 Classification algorithm

The output from the testing phase gave us images obtained in gray-scale. These black and white images were the result of the preliminary segmentation which did not characterize the tumor localization or provide any information regarding the grade of the tumor. To obtain a color-based contrast segmentation result from this initial output, we integrated a separate program with our DenseNet architecture.

4.6 Results

4.6.1 *Training results*

The results obtained from training are displayed in the tables below. Table 4.1 shows the mean of patient loss and patient accuracy. Table 4.2 provides the quantity of different parameters in detail from patient 17 to 20.

Table 4.1   Average value of patient loss and patient accuracy

Sl.No	Patient	Avg.Patient Loss	Avg.Patient Acc
1	17	0.7720	85.3435
2	18	0.4714	91.0136
3	19	0.7456	87.2426
4	20	0.2356	96.4125

Table 4.2   Values obtained for training

Patient	Iteration	Loss	Acc.FLAIR %	Acc.t1 %
17	135	1.4184	78.24	53.50
17	136	0.7730	83.54	87.30
17	137	0.1062	100	100
17	138	1.1066	87.73	73.55
17	139	0.3954	90.91	93.23
17	140	0.6942	89.38	91.61
17	141	1.4753	81.45	55.76
17	142	0.8252	93.29	80.84
17	143	0.0819	100	100
17	144	0.6278	94.27	80.15
17	145	1.0741	84.69	78.76
17	146	1.4878	80.09	68.43
17	147	1.0662	77.89	74.91
17	148	0.6261	92.82	90.05
17	149	0.5661	88.19	85.62
17	150	0.8604	80.41	85.13
17	151	0.4837	90.89	93.63
17	152	0.2011	97.14	98.58
17	153	1.8998	85.24	40.68
17	154	1.5867	62.33	78.56
17	155	0.5730	94.42	83.97
18	1	0.5618	96.01	84.87
18	2	1.6947	90.05	37.15
18	3	1.2617	94.16	54.80
18	4	0.9922	93.32	65.39
18	5	0.7631	94.42	68.49
18	6	0.4781	90.02	89.47
18	7	0.1919	97.71	99.48
18	8	0.1763	99.62	99.39
18	9	0.6648	89.90	79.57
18	10	0.7703	91.12	81.74
		.		
		.		
		.		
18	150	0.3837	94.91	90.51
18	151	0.3845	96.47	84.87
18	152	0.6000	93.61	81.19
18	153	0.5776	87.53	88.80
18	154	0.3697	94.65	90.71
18	155	0.2088	97.48	96.41

Table 4.2 Continued

Patient	Iteration	Loss	Acc.FLAIR t2 %	Acc.t1 %
19	1	1.3867	93.29	61.75
19	2	0.2640	98.99	99.68
19	3	2.4233	89.93	42.01
19	4	0.7689	99.19	67.25
19	5	0.3828	94.18	88.86
19	6	1.3679	86.40	69.73
19	7	0.9030	100	100
19	8	0.8523	98.70	62.53
19	9	0.5055	91.75	84.90
19	10	0.4638	93.00	90.97
		.		
		.		
		.		
19	150	0.0995	100	100
19	151	0.5108	95.05	88.77
19	152	0.2882	100	100
19	153	0.2435	97.40	95.49
19	154	0.3244	100	84.52
19	155	0.4355	94.68	87.44
20	1	1.1837	96.99	66.06
20	2	0.1956	99.83	100
20	3	1.0576	91.00	72.97
20	4	2.2987	89.44	38.51
20	5	1.8808	100	50
20	6	0.9994	84.03	75.84
20	7	1.7880	99.94	50.00
20	8	0.2220	99.68	99.91
20	9	0.7065	92.94	83.88
20	10	0.3926	100	74.28
		.		
		.		
		.		
20	150	0.2136	98.78	95.17
20	151	0.1916	97.74	96.24
20	152	0.1600	100	100
20	153	0.0291	99.68	99.59
20	154	0.0883	100	99.62
20	155	0.2620	95.30	95.52

Note: First column represents the value of iteration; second column shows the value of loss (1-dice score); third column is FLAIR accuracy percentage; and fourth column is t1 percentage accuracy.

Epoch: The data that is used in training the neural network is made to pass through the network multiple times to reach maximum accuracy. Finally, the weights change and the neural network curve changes from underfitting to overfitting.

In this model, we have taken MRI scans of 20 patients. This data is sent through the network five times, i.e., 0–4 epochs. In the above table, we have reported the values of the fourth epoch.

Patient: 20 patients were considered for training this model.

Iterations: When the data for training the neural network is too big, the data is divided into smaller parts called batches. The number of batches needed to complete one epoch is called iterations. In this training model, the data is divided into 155 batches and iterated 155 times.

Loss: Loss is defined as 1-Dice score of each of the iterations. Dice score quantifies the performance of the algorithm.

Acc.FLAIR *t2*%: FLAIR is expanded as fluid attenuated inversion recovery. They are images with suppressed cerebrospinal fluid effects, and they are devoid of the noise the fluid might cause during the scan.

The BraTS dataset is divided into different modalities, and FLAIR *t2* is one of the modalities. This modality has FLAIR images, which specifies the position of the tumor.

Therefore, the accuracy percent of these images is important for increasing the efficiency of the neural network.

Acc.*t1*%: This is one of the modalities of the BraTS dataset. This value specifies the accuracy percent of contrast enhancement. These are the images of the whole tumor and their core.

The results from the training set showed a relatively high patient-wise accuracy (Tables 4.1 and 4.2).

4.6.2 Test results

The visual information extracted from the MRI images had to be provided with in-built multi-class labels through the segmentation algorithm for the testing process. The evaluation of the sensitivity, accuracy, and positive prediction analysis was done by categorizing the various structures that are found to be mutually inclusive gliomal regions.

Tumor region annotations used for the segmentation process validation are as follows:

- Whole tumor region – Denoted as "whole"; this categorizes all four tumor structures
- Tumor core region – Denoted as "core"; this region excludes the part with the edema
- The active tumor region – Denoted as "enhance"; this contains the parts that are specific to the high grade gliomas

The predictive validity of the segmentation test is calculated in a way that corresponds to a certain definitive indicator which is typically referred to as a gold standard.

Enhanced tumor (ET): The areas that show hyper-intensity in T1-Gd when it is compared both with T1 and healthy white matter is defined as enhanced. Regions where there is a leakage due to disrupted blood–brain barrier are where ET is biologically found. This is mostly seen in high grade gliomas.

Whole: The whole tumor is described as the union of non-enhanced tumor, enhanced tumor, and peritumoral edema.

Core: The core tumor is described as the union of enhanced tumor and non-enhanced tumor.

Table 4.3   Values of Dice whole, Dice core, and Dice enhance

Sl.No	Dice Whole	Dice Core	Dice ET
1	0.9960305779569892	0.997740479390681	1.0
2	0.9930840053763441	0.9951382168458781	1.0
3	0.9786905241935484	0.9795201612903226	0.98997020609319
4	0.9961133512544803	0.9988175403225806	1.0
5	0.9956051747311828	0.999907482078853	0.999922603046595
6	0.9839754704301076	0.9975580197132616	1.0
7	0.98791845878 1362	0.9933440860215054	1.0
8	0.9966316084229391	0.9996747311827957	0.9998361335125449
9	0.9753266129032258	0.9992920026881721	0.9995424507168459
10	0.975552979390681	0.9935840053763441	0.9999858870967742
11	0.997167002688172	0.9987939068100359	0.9999419802867383
12	0.9904658378136201	0.9985877016129032	0.9992560483870968
13	0.9785453629032258	0.9952661290322581	1.0
14	0.9834467965949821	0.9894483646953405	0.9999964157706093
15	0.9818430779569892	0.9899939516129033	0.998076164874552

Table 4.3 Continued

Sl.No	Dice Whole	Dice Core	Dice ET
16	0.9773267249103943	0.9828504704301075	0.9992081093189964
17	0.9973293010752688	0.9976474014336918	1.0
18	0.9948823924731183	0.9983751120071684	0.9999894713261649
19	0.9907242383512544	0.995953853046595	1.0
20	0.9859519489247311	0.9919772625448029	1.0
21	0.9913732078853047	0.9939707661290322	0.9995726926523297
22	0.9797576164874552	0.989123767921147	1.0
23	0.9893090277777777	0.9924935035842294	0.9998799283154122
24	0.9780263216845878	0.9841258960573477	0.9996010304659498
25	0.9913753360215054	0.9958382616487456	0.9979521729390681
26	0.9916144713261649	0.9972542562724014	1.0
27	0.9959844310035842	0.9971776433691756	0.9998517025089606
28	0.9870500672043011	0.9885278897849462	0.9997578405017921
29	0.9945105286738352	0.9966945564516129	0.9969462365591398
30	0.9873108198924732	0.9907964829749104	0.9991472894265233
31	0.9958035394265233	0.9988429659498208	0.9990638440860216
32	0.9901963485663082	0.9930421146953405	0.9999866711469534
33	0.9868958333333333	0.9906358646953405	1.0
34	0.9926816756272402	0.9967960349462366	1.0
35	0.9976786514336917	0.9992720654121864	0.9995750448028674
36	0.976233646953405	0.9820928539426523	0.999792226702509
37	0.9859595654121864	0.9859595654121864	0.9998662634408603
38	0.9923506944444445	0.9943670474910394	1.0
39	0.9944155465949821	0.9987927867383513	0.9997655689964158
40	0.9740033602150537	0.988271729390681	0.9994764784946236

4.6.3 Dice scores

The Dice values obtained from our testing output for the different regions are as given in Table 4.3. The scores we obtained come very close to the ideal value of 1. This displays a very high performance index in the segmentation ability of our neural network.

The mean values of Dice scores were calculated to plot boxplot graphs (Table 4.4 and Figure 4.3).

Table 4.4 Mean Dice values

Mean Dice Whole	Mean Dice Core	Mean Dice Enhance
0.988285534274194	0.9936886732750896	0.9993990115367384

4.6.4 Positive predictive value (PPV)

The mean scores obtained for the different image sets are as given in Table 4.5. The high PPV is an indicator of the certainty of the output of our model and the large utility of its patient accuracy.

The mean values of PVV scores were calculated to plot boxplot graphs (Table 4.5 and Figure 4.3).

4.6.5 Sensitivity

The mean scores we obtained on our sensitivity calculations on the different image sets are as given in Table 4.6. The high sensitivity of our DenseNet architecture is a good measure of the percentage of the afflicted people correctly diagnosed.

The mean values of sensitivity scores were calculated to plot boxplot graphs (Table 4.6 and Figure 4.3).

Table 4.5   Mean PPV (positive predictive value) values

Mean PPV Whole	Mean PPV Core	Mean PPV Enhance
0.9882285534274194	0.9936886732750896	0.9993990115367384

Table 4.6   Mean sensitivity values

Mean Sensitivity Whole	Mean Sensitivity Core	Mean Sensitivity Enhance
0.9882285534274194	0.9936886732750896	0.9993990115367384

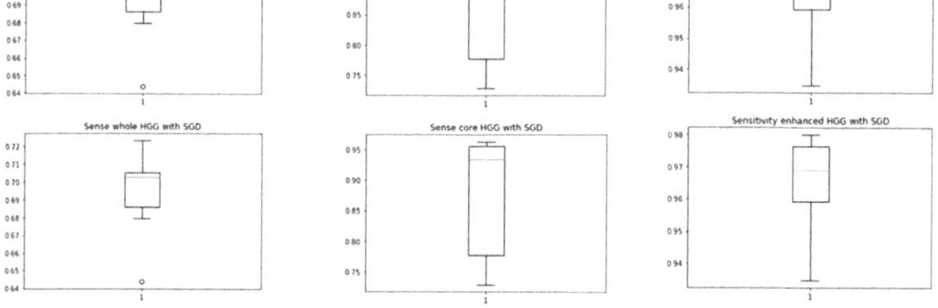

Figure 4.3   Box plot of Dice (whole, core, and enhance), PPV (whole, core, and enhance), and sensitivity (whole, core, and enhance) vs. mean Dice, mean PPV, and mean sensitivity, respectively.

Conclusion

A field of study is often constrained or allowed to progress by the advancements of the tools employed. This makes the development and polishing of new technologies a primary area of research which has led to the congregation of several experts to evaluate and design novel techniques. Deep learning has shown a high level of versatility that has allowed the integration of this rapidly developing field into various other more traditional areas of research. Medical science is one such domain that has employed deep learning tools for the better execution of diagnosis, information storage, and predictive capabilities offered by the largely self-evolving and intuitive design of deep learning products. In the medical field, where previous niche knowledge is a criterion for future assumptions and deductions, neural networks offer the optimum solution for a range of different problems.

Gliomas have been a prevalent condition in a large part of the human population and statistics have indicated that these numbers are always on the rise. The diagnosis and treatment of this largely fatal condition requires a tool that can operate at high accuracy while holding up the time sensitive nature of the issue. Neural network designs are created for these operations to accommodate the errors that can be generated by the manual approach and to continuously learn and develop its prediction and diagnostic abilities based on previous data.

Our usage of the DenseNet architecture in the segmentation of MRI glioma images gave us results that show a strong backing for the performance and utility of neural networks in the area of medical diagnosis. The Dice score values obtained from our network training process display segmentation abilities of very high accuracy which matched some other recent neural network architectures designed and used in recent studies and surpassed many others that existed in the past few years. Our sensitivity values showed a very low tendency for error and high confidence levels which is a very important criterion while handling the diagnosis of critical conditions such as tumors where even tiny fallacies can result in fatality or irreversible damage. The utility of our architecture is further enhanced by its strong positive prediction rates which offer a constant comparison to the disease prevalence in the general population in order to offer a good score of the diagnostic certainty. The WHO criteria for categorization of the severity of the gliomas are another factor that we have integrated into our algorithm by training the networks separately with high grade and low grade glioma images. Localization of tumors is also defined with highly enhanced features to better aid in surgical procedures.

Typically, neural networks require a large number of initial training data for optimal functioning. The DenseNet architecture we used has the added advantage of bypassing this issue by the collective reuse and processing of the output feature maps. Our network was efficiently trained using the MRI image set from just 20 patients. We have provided for the visualization of different segments and regions of varied tissue conditions in our program. The training and testing of our architecture was done with MRI images with T1, T2, and FLAIR contrast images as a while to obtain and extract accurate features in the segmentation procedure. We annotated different tumor structures with appropriate RGB color values and evaluated a frame-by-frame output to get a model of the severity and localization of the glioma in the brain. The interactive software program we used to observe this segmented output allows an easy-to-use interface by doctors for analysis of the patient's condition.

Future Prospects

Neural network implementation in medical diagnosis has come a long way since its first elucidation by Miller in his review in 1992 [16]. Even with the progress of the field, there is a necessity to identify its drawbacks in order to make room for researchers in the future to address these problems.

The feature exaggeration offered by DenseNets is also subject to erroneous extraction due to the use of dense skip connection. The model cannot potentially utilize all the parameters that could be amended by using the features on other convoluted network architecture. An architecture can be designed to integrate these components or even design a novel approach based on previous architectures.

Another potential problem with the use of these models is that these products are utilized by clinicians who are unskilled with respect to the use of advanced deep learning tools. An interface that can allow them to conveniently access and store files and perform various operations on them is a necessity to actively involve neural networking at the clinical level. Easy-to-use platforms that have networks trained to segment and classify images at the back end have to be designed for increased usage capability.

Our architecture was trained using MRI images as they are one of the most widely used imaging techniques for gliomas. But MRI images are known to produce images with non-uniform intensity which can manifest as varying changes in the pixel values of the images produced. Therefore, the architecture can be modified to accommodate image files produced by other high end imaging tools such as PET and CT. The DenseNet architecture implemented in this project can also be trained to segment other tumors by training them on the particular tumor image set. This can diversify the role of the neural net in different aspects of the diagnostic process.

There can still be modifications made to further classify particular sections of the gliomas beyond the basic high grade and low grade division. It is also necessary to integrate neural network structures that can process glioma images obtained from other imaging tools apart from MRI devices to offer a state-of-the-art resource for medical professions to work on with minimal skill specialization required.

References

[1] Omuro A, DeAngelis LM. 'Glioblastoma and Other Malignant Gliomas: A Clinical Review.' JAMA. ;310(17) [2013] 1842–1850. doi:10.1001/jama.2013.280319

[2] Louis DN, Holland EC, Cairncross JG. 'Glioma classification: a molecular reappraisal.' The American journal of pathology vol. 159,3 [2001]: 779-86. doi:10.1016/S0002-9440(10)61750-6

[3] Ohgaki H., Kleihues P. 'Epidemiology and etiology of gliomas.' Acta Neuropathol [2005] 109, 93–108 https://doi.org/10.1007/s00401-005-0991-y

[4] Li Sun , Songtao Zhang , Hang Chen and Lin Luo 'Brain Tumor Segmentation and Survival Prediction Using Multimodal MRI Scans With Deep Learning.' [2015]

[5] Yue Cao, Pia C. Sundgren, Christina I. Tsien, Thomas T. Chenevert, Larry Junck 'Physiologic and Metabolic Magnetic Resonance Imaging in Gliomas' Journal of Clinical Oncology [2006] 24:8, 1228-1235

[6] Villanueva-Meyer JE, Mabray MC, Cha S, 'Current Clinical Brain Tumor Imaging', Neurosurgery, 81,[2017], 397-415

[7] El-Sayed A. El-Dahshan, Heba M. Mohsen, Kenneth Revett, Abdel-Badeeh M. Salem, 'Computer-aided diagnosis of human brain tumor through MRI: A survey and a new algorithm' Expert Systems with Application [2014], 5525-5545

[8]   Riccardo Miotto, Fei Wang, Shuang Wang, Xiaoqian Jiang, Joel T Dudley 'Deep learning for healthcare: review, opportunities and challenges' Briefings in Bioinformatics, Volume 19, Issue 6, [2018] Pages 1236–1246, https://doi.org/10.1093/bib/bbx044

[9]   Lele Chen, Yue Wu Adora, M. DSouza , Anas Z. Abidin , Axel Wismüller, Chenliang Xu, 'MRI Tumor Segmentation with Densely Connected 3D CNN', University of Rochester, New York [2017]

[10]   Shi Z, He L, Suzuki K, Nakamura T, Itoh H. 'Survey on Neural Networks Used for Medical Image Processing.' International journal of computational science vol. 3,1 [2009] 86-100.

[11]   Zaharchuk G, Gong E, Wintermark M, Rubin D, Langlotz CP, 'Deep Learning in Neuroradiology' American Journal of Neuroradiology, [2018]

[12]   J´egou1 S., Drozdzal M., Vazquez D., Romero A., Bengio Y. 'The One Hundred Layers Tiramisu: Fully Convolutional DenseNets for Semantic Segmentation' Computer Vision Center, Barcelona. [2017]

[13]   Sérgio Pereira, Adriano Pinto, Victor Alves, and Carlos A. Silva, 'Brain Tumor segmentation using convolutional neural networks for MRI images' IEEE TRANSACTIONS ON MEDICAL IMAGING, VOL. 35, NO. 5, [2016]

[14]   Zhao X., Wu Y., Song G., Li Z., Fan Y., Zhang Y., 'Brain tumor segmentation using a fully convolutional neural network with conditional random fields' MICCAI, Athens, Greece, October 17, 2016, 75{87 [2016].

[15]   Kamnitsas, K., Ledig, C., Newcombe, V. F. J., Simpson, J. P., Kane, A. D., Menon, D. K., Rueckert, D., Glocker, B., 'Efficient multi-scale 3d CNN with fully connected CRF for accurate brain lesion segmentation' Medical Image Analysis 36, 61{78 [2017].

[16]   Trevethan, Robert. "Sensitivity, Specificity, and Predictive Values: Foundations, Pliabilities, and Pitfalls in Research and Practice." Frontiers in public health vol. 5 307. 20 Nov. 2017, doi:10.3389/fpubh.2017.00307

[17]   Miller AS, Blott BH, Hames TK 'Review of neural network applications in medical imaging and signal processing' .Med Biol Eng Comput. [1992] Sep; 30(5):449-64.

[18] Gao Huang, Zhaung Liu, Killian Q Weinberger, Laurens van der Maaten, "Densely connected convolutional networks", IEEE EXPLORE – [2018]

[19] Jean Stawiaski1, "A pre-trained Dense-net encoder for brain tumor segmentation.", 19 Nov 2018

[20] Ashia C. Wilson, Rebecca Roelofs, Mitchell Stern, Nathan Srebroy, Benjamin Recht, "The Marginal Value of Adaptive Gradient Methods in Machine Learning.", University of California Berkeley [2016]

[21] Hao Dong, Guang Yang, Fangde Liu, Yuanhan Mo, Yike Guo, "Automatic Brain Tumor Detection and Segmentation Using U-Net Based Fully Convolutional Networks", Data Science Journal of London-2017.

[22] B. Menze et al., "The multimodal brain tumor image segmentation benchmark (BRATS)," IEEE Trans. Med. Imag., vol. 34, no. 10, [2015] pp.1993–2024,.

[23] Martin Kolarˇík , Radim Burget , Václav Uher , Kamil R íha, Malay Kishore Dut, "Optimized high resolution 3D Dense-U-NET network for brain and spine segmentation", MDPI applied sciences January 2019.

5

Automatic Liver Tumor Segmentation from Computed Tomography Images Based on 2D and 3D Deep Neural Networks

William Tustumi, Guilherme P. Telles, and Helio Pedrini

Institute of Computing, University of Campinas,
Campinas-SP 13083-852, Brazil

Abstract

Segmentation of liver tumors is the process of voxel classification between tumor and healthy tissue performed from a volume of computed tomography. The increasing quality of medical image acquisition methods has allowed the identification, location, and diagnosis of diseases, avoiding very intrusive surgeries. This analysis is vital to decide the most appropriate treatment for the patient. In particular, the segmentation of the tumors is used to decide the feasibility of extracting the tumor and helps to specify the operative plan. The segmentation process of regions affected by tumors, when performed manually, requires time and experience from medical specialists, as it involves creating a tumor mask for each of the slices of the tomography. This task is particularly challenging when patients are located in poorly served regions and away from specialized medical services. Two types of approaches have traditionally been proposed to speed up and facilitate the segmentation of CT scans, one completely automatic and the other based on human intervention. In this work, we focus on fully automatic techniques for segmenting liver tumors. Despite the convolutional neural networks achieving significant results in the areas of image segmentation and classification, the segmentation of tomographic volumes presents new challenges, such as the introduction of a dimension of spatial relationships, artifacts from the extraction of images and a limited number of examples for training. In our research, we investigated the trade-off between computational resources and segmentation quality. Initially, we analyzed the performance of several convolutional networks and tested layers of different networks following the Effcient Net balancing model. Next, we expanded the layers for three-dimensional convolutions and tested layers that handled the dimensions of the volume separately. Finally, we evaluated the execution time of our models in equipment with limited processing and memory. Although our models have obtained inferior results in terms of effectiveness when compared to other methods in the literature, their execution proved to be viable in a restrictive computational environment. The experiments were performed on the liver tumor segmentation challenge (LiTS) database.

5.1 Introduction

The technology for medical image acquisition has advanced significantly over the last few decades, transitioning from film images to digital images, which allows for higher resolution and three-dimensional (3D) imagery. A number of methods that allow discriminating internal tissues have been developed over the years, including computed tomography (CT), chest X-Ray, magnetic resonance imaging and ultrasonography.

Cancer was the first or second leading cause of death in most countries over the world in 2020, with approximately 10 million deaths. Although liver cancer is not among the five most common types of cancer, it is the third most deadly, with 830,000 estimated deaths (Sung *et al.* [2020]).

Due to its impact on global health, a substantial amount of investment has been directed towards the treatment, diagnosis, and detection of this type of disease. However, there is a disparity between the resources available in wealthy and poor countries, that is, while 90% of the high-income countries reported availability of treatment services, less than 30% of low-income countries have these services. In addition, approximately 70% of cancer deaths occur in low- and middle-income countries.

Often, primary tumors are located in the abdomen and metastasize to the liver. Therefore, the liver and liver lesions are routinely analyzed when breast, pancreas, and colon cancer are diagnosed. The results of the segmentation of liver tumor images are vital for planning, diagnosis, and monitoring the treatment response time. It is also an important resource for treatments such as thermal percutaneous ablation (Rossi *et al.* [1996]), percutaneous ethanol injection (Livraghi *et al.* [1995]), and radiotherapy surgical resection (Albain *et al.* [2009]), among others. However, the number of available specialists is insufficient to cover all demanding patients, especially in isolated areas and underdeveloped countries. Furthermore, the manual segmentation process of a CT scan is costly, time consuming, hard to reproduce, and heavily dependent on the human specialist (Bilic *et al.* [2019]).

As a consequence of such limitations on the segmentation process, the labeled datasets available for training machine learning models for this problem are relatively poor when compared to datasets used in broader image segmentation problems, where deep learning models have already surpassed human accuracy. In addition to limited datasets, the region of interest in each CT scan is a fraction of the total volume, which creates an unbalanced class distribution that can bias classification models.

This work aims at developing an automatic method for liver tumor segmentation from 3D CT images using deep neural networks (DNNs). Promising results were already achieved with DNNs in skin lesion segmentation (Tang *et al.* [2019], Tschandl *et al.* [2019], Xie *et al.* [2020]), brain tumor segmentation (Chen *et al.* [2020, 2019], Wang *et al.* [2019a]), and also liver tumor segmentation (Han [2017], Li *et al.* [2018], Wang *et al.* [2019b]). We use the architecture proposed by Han [2017] as a baseline and then investigate and extend the applicability of new methods, such as depth and width balance (Tan and Le [2019]), 3D-convolutional layers (Milletari *et al.* [2016]), separable convolutions layers (Howard *et al.* [2017]), and cascade architecture (Li *et al.* [2017]).

This text is organized as follows. Sections 5.2 and 5.3 describe relevant concepts and works related to the topic under investigation. Section 5.4 presents the methodology proposed in this work. Section 5.5 describes and discusses our experimental results. Section 5.6 contains our concluding remarks.

5.2 Related Concepts

In this section, we briefly describe basic concepts related to computed tomographic volume acquisition, image preprocessing and segmentation, data augmentation, and neural networks, which are necessary to understand the subsequent sections of the research work.

5.2.1 Segmentation

Image segmentation is the process of delimiting object boundaries in an image (Gonzalez *et al.* [2004], Le *et al.* [2021], Meng *et al.* [2021]). The segmentation problem is an expansion of the localization and detection problems since their responses can be extracted directly from the segmentation solution.

Segmentation typically combines different image features to achieve a successful solution. Many segmentation algorithms search for pattern discontinuations and pixel similarities to find object boundaries. However, most of them are dependent on the image type and characteristics, which makes it difficult to reuse solutions from different problems.

The concept of image segmentation can be extended to volume segmentation, where the object delimitation is not defined by the boundaries of pixels but by the boundaries of voxels.

5.2.2 Computed tomography

Hounsfield [1973] developed the computed tomography (CT) scanning technology. The process generates a faithful 3D distribution of X-ray attenuation values per volume unit, which generates the inverse Radon transform from projections in an axial slice. Modern scanners that use multi-slice technology can achieve less than 1 mm per slice depending on the level of detail needed for the region of interest (Toennies [2017]), which is far better than the slice granularity achieved in the early days of CT when the thickness was normally above 5 mm.

The attenuation values from the CT volume are normalized to the Hounsfield unit (HU) scale to make them independent from the X-ray energy output. The normalization is expressed in eqn (5.1).

$$HU(\mu) = 1000 \frac{\mu - \mu_{water}}{\mu_{water} - \mu_{air}} \qquad (5.1)$$

where μ_{air} and μ_{water} are the linear attenuation coefficients of air and water, respectively.

The values of HU usually range from −1000 to 3000. The attenuation values of air, fat, and water are considerably different, whereas the difference between distinct soft tissues is minimal, which makes their classification more difficult. Table 5.1 shows the HU values for different materials and body tissues.

Table 5.1 Values of Hounsfield unit (HU) for different materials and body tissues. Values extracted from Toennies [2017]

Air	Fat	Water	Blood	Muscle	White Matter	Grey Matter	Bone
−1000	−100	0	30–45	40	20–30	37–45	>150

5.2.3 3D convolution

Convolution is the process of adding each pixel of an image to its neighbors weighted by a convolution matrix (kernel). 3D convolutions are a natural expansion of 2D convolutions for volumetric data. While 2D-convolution filters have four dimensions (height, width, channel input, and channel output), 3D convolution filters have five dimensions, where an additional depth dimension is added after the width. This new dimension adds a z-axis to the filter movement. The added dimension makes it straightforward to capture the relationship between features on the newly created axis, but it also increases the computational cost of the model. Therefore, many models opt to use a 2.5D approach: volumetric data with 2D convolutions, where the depth dimension is mapped to the input filters. Figure 5.1 depicts the differences between 2D convolution and 3D convolution.

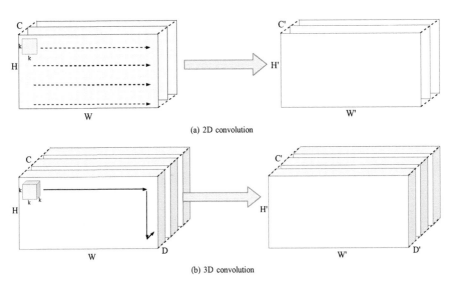

(a) 2D convolution

(b) 3D convolution

Figure 5.1 Example of (a) 2D convolution and (b) 3D convolution diagrams. The traced arrows in the 2D convolution show the filter movement through the input. The filter in a 3D convolution must follow a similar path from the 2D convolution. However, since it has an additional dimension to move, it repeats the 2D- movement for each entry on the z-axis. Figure 5.1(b) indicates the dimensions by which the filter will move without the complete path to avoid cluttering the image. H, W, D, and C denote the input: width, height, depth and channels, respectively, while H′, W′, D′, and C′ are their output counterparts.

5.2.4 Separable convolution

Separable convolutions are divided into two groups: spatial and depthwise. The first occurs in the spatial dimension; in the case of images, it separates width and height of the kernel, as in Figure 5.2. The second type separates the convolution in spatial aggregation and feature combination; in the case of images, the first filter has its original width and height, whereas the second filter is a 1st filter that transforms the number of features into the desired output. Figure 5.3 illustrates the convolution process.

The depthwise separable convolutions are the basis for more efficient neural network architectures, because they reduce the number of computations and parameters without decreasing the descriptive power. This is not the case

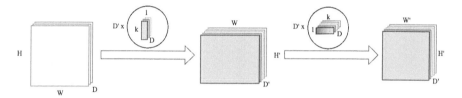

Figure 5.2   Diagram of a spatial separable convolution. The values on the left side of the circle represent how many repetitions of the kernel are used in the convolution.

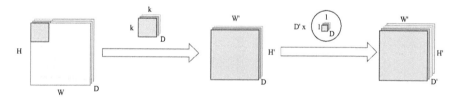

Figure 5.3   Diagram of a depthwise separable convolution. The first convolution separates the input into D layers, where each layer is convoluted by a layer of the kernel. In the image, this operation is represented by its colors. The value on the left side of the circle represents how many repetitions of the kernel are used in the pointwise convolution.

Table 5.2   Expressions for the number of parameters from different 2D convolutions, where k is the size of the kernel, c_i is the number of input channels, and c_o is the number of output channels

Convolution	Number of parameters
Conventional	$k^2 * c^i * c^o$
Spatial separable	$2 * k * c^i * c^o$
Depthwise separable	$k^2 * c^i + c^i * c^o$

with spatial separable convolutions, where the only type of kernels they can represent are those composed of separable filters.

The depthwise convolution separates the convolution into a group convolution and a pointwise convolution. The group convolution maintains the number of feature maps and transforms the spatial dimension. The pointwise convolution uses a 1se kernel with stride 1 that transforms the number of channels but maintains the spatial dimension.

From Table 5.2, it is possible to conclude that the conventional convolution scales worse and the depthwise convolution scales better than the spatial convolution for values small values of k, which is the usual.

5.2.5 Depthwise spatio-temporal separate

The depthwise spatio-temporal separate (DSTS) module, proposed by Zhang *et al.* [2019a], separates 3D convolutions into two branches: temporal and spatial. Both branches use depthwise convolutions but with different filters.

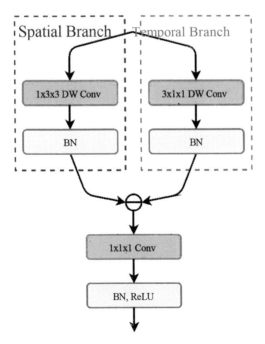

Figure 5.4 Diagram of a depthwise spatio-temporal separate module. In the diagram, BN is a batch normalization module that contributes the network's stability and improves training speed; and ReLU is the activation function rectified linear unit that adds non-linearity to the sequence of convolutions.

The temporal side has filters of size $3 \times 1 \times 1$ and the spatial side uses filters of size $1 \times 3 \times 3$. A pointwise convolution combines the features after a concatenation.

Figure 5.4 shows a diagram of the DSTS module. In addition to forcing the network to learn different relationships at each branch, the separation has the added benefit of reducing the number of parameters and the computational cost. The number of parameters is reduced from $27c^2$ with a normal 3D convolution with kernel size 3 to $c^2 + 12c$, where c is the number of channels in the layer.

5.2.6 U-net

U-Net is a convolutional neural network (CNN) created to segment biomedical images that improved accuracy and training speed of previous segmentation methods by creating the entire segmentation mask instead of classifying small patches of the image at a time. It also benefits heavily from data augmentation explained in Section 5.4.3 and, therefore, can be trained on a small dataset (Ronneberger *et al.* [2015]).

Figure 5.5 shows a diagram of the U-Net model. The left side of the architecture shrinks the image dimensions and increases its depth (encoder), whereas the right side expands the image to create the segmentation map (decoder). The horizontal blue arrows represent shortcuts on the model, which combine previous layers from the encoder to the decoder. The layers

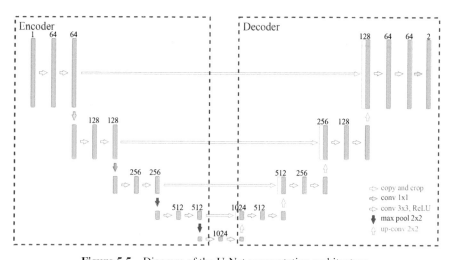

Figure 5.5 Diagram of the U-Net segmentation architecture.

are combined using a concatenation operator that simply append the feature maps along their channels.

5.2.7 EfficientNet

Tan and Le [2019] observed that models that balanced their depth, width, and resolution reached a higher accuracy for the same computational cost than models that favored only one of these coefficients. With that in mind, they proposed a compound scaling method that balances these three coefficients, given a target computational constraint. In their experiments, they chose the number of floating point operations (FLOPs) to determine the model size and used the accuracy on ImageNet to measure its performance.

The baseline coefficients were discovered through a grid search with a target of 400M FLOPs. Notably, the scaling factor of the network depth is linear, while the scaling factor of width and resolution is quadratic. In other words, doubling the network depth doubles the number of FLOPs, while doubling either the width or the resolution quadruples the number of FLOPs. The base model uses the mobile inverted bottleneck from Howard *et al.* [2017] coupled with squeeze-and-excitation module from Hu *et al.* [2018] as a building block.

5.2.8 Loss function

A loss function guides how the weights within the learning method are updated. It can be formulated either from the difference between expected output and predicted output or based on a desired feature that the output must present. The former can only be used in supervised training, whereas the latter is used in unsupervised training or as a complementary loss to a supervised training. The most common loss functions used for image segmentation architectures are weighted cross entropy (WCE), Dice distance, and Jaccard distance.

The Dice distance and Jaccard distance are very similar, as both originate from metrics used to measure the segmentation quality. The Dice distance is based on the Sorensen–Dice coefficient, whereas the Jaccard distance is based on the Jaccard index. The Dice coefficient is equal to twice the size of the intersection between sets Y and Y divided by the sum of the elements on both sets:

$$\text{Dice} = \frac{2 \cdot |Y \cap \hat{Y}|}{|Y| + |\hat{Y}|}$$

In order to use the Dice equation as a loss function for supervised training of a neural network, it is necessary to adapt its formula, since the function was proposed to work with discrete data. The following equation is the Dice loss function adapted for a minimization problem that allows gradient backpropagation:

$$\text{Dice distance} = 1 - \frac{2 \cdot \sum_{i=1}^{N} y_i \cdot \hat{y}_i}{\sum_{i=1}^{N} y_i^2 + \hat{y}_i^2}$$

5.2.9 Metrics

We present in the following sections the metrics used to evaluate the segmentation performance; those metrics were chosen because they are standard on image segmentation and were used for performance comparison on the Liver tumor segmentation competition (LiTS).

5.2.9.1 *Overlapping metrics*

Overlapping-based methods can determine whether voxels from the predicted segmentation appear in the ground-truth segmentation. On the other hand, they cannot calculate the distance between the voxels that are present in one set and not in the other.

Dice similarity coefficient: Dice [1945] proposed a metric to quantify the matches between two sets. Many other researchers have adopted the metric to evaluate the performance of medical segmentation. The Dice metric compares the overlapping ratio between the predicted mask and a manual segmentation done by a specialist.

Given two sets of voxels A and B, where A denotes a predicted segmentation of liver tumors and B denotes the ground-truth of liver tumors, we can calculate the Dice similarity coefficient through eqn (5.2), where $|A \cap B|$ is the intersection size between the two sets, and $|A|$ and $|B|$ are the sizes of the respective sets:

$$\text{Dice}(A, B) = \frac{2 \cdot |A \cap B|}{|A| + |B|} \tag{5.2}$$

Jaccard volume overlap error: The Jaccard similarity coefficient (Jaccard [1901]) of two sets is defined as their intersection size over their union size.

The Jaccard volume overlap error (VOE) is the complement of the Jaccard coefficient. Using A and B as the same pair of sets of the Dice similarity coefficient, VOE can be written in terms of A and B, as shown in eqn (5.3):

$$\text{VOE} = 1 - \frac{|A \cap B|}{|A \cup B|} \tag{5.3}$$

It is possible to observe that VOE measures the difference between two sets, so that similar sets have low VOE values, which is the opposite of the Dice similarity coefficient.

Surface distance metrics: Before describing a number of surface-distance-based metrics, it is important to introduce some important concepts. Let x be the distance between a voxel and a set of voxels Y, defined as

$$\text{distance}(x, Y) = \min_{\forall y \in Y} \text{distance}(x, y) \tag{5.4}$$

The function distance in eqn (5.4) is the Euclidean distance between the coordinates of voxels x and y. Let B_A be the set of all voxels in A's border and B_B denote the set of all voxels in the B's border.

Metrics based on surface distance consider only voxels that belong to the segmentation surface. This char acteristic can mask large volume errors that do not deform the segmentation surface.

Average symmetric surface distance: The average symmetric surface distance (ASSD) metric averages the distance between all voxels in B_A to B_B and all voxels in B_B to B_A. This metric gives a general idea of how close the borders are in their entire extension. However, due to its robustness to outliers, it masks any isolated δ mistakes. Since a variation δ over a single distance has the same effect as a variation of $\frac{\delta}{|B_A|+|B_B|}$ spread over all distances, consequently, a large surface has a smaller variation in ASSD value:

$$\text{ASSD} = \frac{1}{|B_A|+|B_B|} \cdot \left[\sum_{x \in B_A} \text{distance}(x, B_B) + \sum_{y \in B_B} \text{distance}(y, B_A) \right] \tag{5.5}$$

Maximum symmetric surface distance: As the ASSD metric cannot detect rare segmentation errors, the maximum symmetric surface distance (MSSD) is sensitive to them, which makes them complementary metrics:

$$\text{MSSD} = \max_{\forall x \in B_A} \left(\text{distance}(x, B_B) \right) \tag{5.6}$$

5.3 Related Work

Han [2017] won the LiTS challenge at the 2017 IEEE International Symposium on Biomedical Imaging. The proposed solution uses two fully convolutional neural networks (FCNNs) with long-range shortcut connections from U-Net (Ronneberger *et al.* [2015]) and short-residual connections from ResNet (He *et al.* [2016]). The first network creates a coarse segmentation of the liver, which is used to limit the number of slices employed in the second network to the slices within the liver segmentation. The second network then creates a refined segmentation of the liver and the lesion segmentation.

The author decided to use 2D convolution to reduce processing and memory resources to leverage the z-axis relation among voxels. Five slices are used to predict the in-between slice. This method is called 2.5D convolution network, since it still considers volume relations but uses 2D filters. The authors addressed the difference in spacing between slices by undersampling slices so that each volume had a spacing of 1 mm × 1 mm × 2.5 mm. The undersampling method was only used in the first network, whereas the second network receives the input with the original spacing, since crucial information could be lost with the undersampling.

The networks were trained using stochastic gradient descent through 50 epochs and took 4 days for each network on a single NVIDIA TitanX GPU. Each prediction slice required 0.2 s to be processed. The code was implemented in Caffe package. The model had 24M parameters. The solution obtained a global Dice coefficient of 0.67, volume overlap error (VOE) score of 0.45, average symmetric surface distance (ASSD) of 6.660, and maximum symmetric surface distance (MSSD) of 57.930 on the LiTS benchmark for lesion tumor segmentation.

Yuan [2017] earned the fifth place in different metrics at the 2017 International Conference on Medical Image Computing and Computer Assisted Intervention (MICCAI) on the LiTS'2017 benchmark. The proposed solution has three steps: liver localization, liver segmentation, and tumor segmentation. The liver localization uses a small fully convolutional neural network (FCNN) with 19 layers and 230,129K parameters. The localization step uses slices of 128 × 128 pixels and undersamples the slices to 3-mm thickness to create a coarse segmentation that is used in the next step.

The liver segmentation step creates a volume of interest by expanding 10 voxels in each direction of the coarse segmentation of the previous step. The sampling frequency is increased, so that the space inter-slice is 2 mm. Moreover, the resolution is expanded to 256 × 256 × 256. The network used in this step is more sophisticated, with 5M parameters and 29 layers.

The tumor segmentation step uses the same network as the liver segmentation stage; however, the original slice spacing is used to avoid missing small lesions due to image blurring. During training, only the slice with tumors were used for training in order to reduce computational time. As postprocessing, a filter was used to discard any tumor outside the predicted liver mask. The final liver tumor mask is the result of a bagging-like ensemble of six models. The loss function used was the Jaccard distance.

The networks were trained using Adam optimizer and required 1.57 days for liver localization and segmen tation and 8.518 days for tumor segmentation on a single NVIDIA GTX 1060 GPU. Each test case took 33 s for prediction on average. The model was implemented in Theano and used 36M parameters. The solution obtained a global Dice coefficient of 0.82, a Dice per case of 0.657, a VOE score of 0.378, an ASSD of 1.151, and an MSSD of 6.269 on the LiTS'2017 benchmark for lesion tumor segmentation.

Li *et al.* [2018] improved the segmentation performance after the end of the 2017 MICCAI challenge on the LiTS'2017 benchmark. The proposed method used traditional 2D and 3D convolutions with a hybrid feature fusion layer to leverage both intra- and inter-slice features. The basic building blocks of the network used densely connected convolutions to gradually expand the number of channels and improve information flow between all closely connected layers. The model created a coarse liver segmentation using a 2D ResNet, which is used as input to the 2D DenseUNet and 3D DenseUNet. Both DenseUNets created feature maps that were combined and fed to the hybrid feature fusion that generates the final liver and tumor segmentation. The adopted loss function was weighted cross entropy as loss function; the class weights were determined through an empirical test.

The networks were trained using stochastic gradient descent and took 21 hours for the 2D DenseUNet and 9 hours for the 3D DenseUNet on 2 NVIDIA Xp GPUs. Each test case took between 30 and 200 s. The code was implemented in TensorFlow and the model has 114M parameters, achieving a global Dice coefficient of 0.82.4 and a Dice per case of 0.72.2 on the LiTS'2017 benchmark.

Zhang *et al.* [2019a] proposed an efficient hybrid convolutional model based on depthwise separable convolu tions and spatial separable convolutions. Initially, they introduced a spatial and temporal factorization module to replace the traditional 3D convolutions. This module is explained in Section 2.5. The model is hybrid since it uses 2D convolutions to extract low-level features and then uses 3D convolutions at deeper levels to reduce com putation, while retaining the spatial connections from the 3D convolutions.

The output block from the encoder is a depthwise spatio-temporal atrous spatial pyramid polling (DST-ASPP) adapted from Chen *et al.* [2017a] and Chen *et al.* [2017b]. The DST-ASPP structure is used to combine multi-range features to calculate the segmentation output. The loss function used was a combination of cross entropy and multi-label Dice.

The network was trained with Adam optimizer and took 24 hours to train on an NVIDIA Tesla P100. Each test case run in 10–80 s. The model has 3.6M parameters and achieved a per case Dice of 0.730 and a global Dice of 0.820 on the LiTS'2017 benchmark.

The model proposed by Wang *et al.* [2019b] is based on the mask R-CNN (He *et al.* [2017]) architecture. This model generates bounding boxes around objects of interest in the image and then performs classification and segmentation in these patches. The main contribution of the work developed by Wang *et al.* [2019b] was the introduction of an attention module to generate the feature maps. The attention module was separated into spatial attention and channel attention. The channel attention is similar to the SE module of Hu *et al.* [2018], whereas the spatial attention is inspired by the work of Wang *et al.* [2018]. The model achieved a per case Dice score of 0.741 and a global Dice score of 0.813 on the LiTS'2017 benchmark. The hardware specification and the time spent to train the model were not mentioned by the work of authors.

5.4 Methodology

Figure 5.6 shows a high-level diagram of our solution. The method is separated into liver segmentation and tumor segmentation. In the second stage, the mask generated in the first stage is used to focus the model on the important areas of the volume. We present a detailed explanation of each stage in Sections 5.4.1–5.4.8.

Figure 5.6 shows a high-level diagram of our solution, which is composed of two phases: (i) the training stage, where we calibrate the network's weight, delimited by the blue dotted rectangle in the diagram, and (ii) the prediction stage, where we create our mask predictions, delimited by the orange dotted rectangle. Our model has two pipelines: one creates liver segmentations and the other creates tumor segmentations. The result of the liver pipeline is combined with the tomographic volume to create the input for the tumor pipeline.

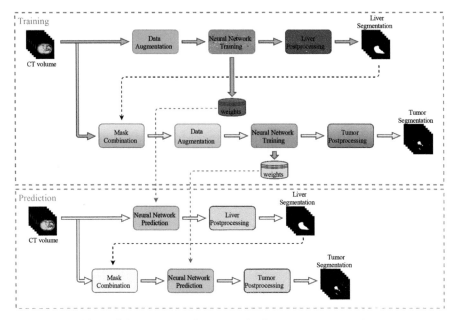

Figure 5.6 Overview diagram of the liver tumor segmentation.

5.4.1 Data normalization and compression

We clamp the HU values of the CT volumes to the [heume 200] range to eliminate noise from materials that are not relevant to lesion or liver segmentation. The clamping operation eliminates all metals, bones, and air from the volume. We convert the values to 32-bit floating point numbers and apply a min–max normalization to limit the values to the range of 0 and 1, which generally improves the model convergence time and avoids floating-point precision errors.

Then, we compress the CT volume and the ground-truth together in an NpzFile to reduce disk space consumption. We also add voxel spacing and volume dimensions to the compressed file as metadata.

5.4.2 Batch sampling

We adopted two different methods to generate samples, one for models assembled with 2D convolutions and the other for models assembled with 3D convolutions. Each sample has two components: axial slices from the CT volume and axial slices from the ground-truth mask.

Consider $CT[s_k]$, the slice with index s_k in the CT volume, GT $[s_k]$, the slice with index s_k in the ground truth, $CT[s_i : s_j]$, the slices with indices from s_i to s_j in the CT volume and $GT[s_i : s_j]$, and the slices with indices from s_i to s_j in the ground truth. Suppose that we are sampling the slice with index s in the CT volume with the first method. Then the CT volume component of the sample is CT $[s_k - w : s_k + w]$ and the ground-truth component is GT $[s_k]$, where $w \in \mathbb{N}+$. If we use the second sampling method for the block of slices $CT[s_i : s_j]$, then the CT volume component of the sample is CT $[s_k - w : s_k + w]$ and the ground-truth component is $GT[s_i : s_j]$ where $w = \dfrac{s_i - s_k}{2}$.

A batch generator samples every slice designated for training once per epoch during training. The sampling method includes the following steps: a random volume is chosen; each slice is used in that volume in a random order; another volume is chosen that was not elected at the current epoch. This method allows proper disk access without compromising the random sampling aspect of the batch generator.

5.4.3 Data augmentation

The data augmentation process is only used at the training stage. We perform data augmentation using geometric and intensity transformations. The types of geometric transformations applied are: rotation, flipping, scaling, translation, shearing, and elastic transformation. We use small random intensity perturbation as our only method of intensity transformation.

We limit the magnitude of the geometric transformations, because the model benefits from positional and structural information. We maintained small intensity disturbances to prevent the voxels from being perceived as different

Table 5.3 Rotation and shearing ranges are in degrees, the translation range is in pixels, random intensity noise range uses the measures of the intensity values after normalization, but the range is less than 2 HU, and the scaling range is measured in percentage of the original image size

Transformation	Range
Rotation	[−15, 15]
Shearing	[−10, 10]
Translation	[−16, 16]
Random intensity noise	[−0.5, 0.5]
Scaling	[−12.5, 12.5]

materials. Table 5.3 shows the parameters used in each transformation. The elastic transformation is performed in only 10% of the images.

5.4.4 Neural network architecture

All architectures used in the experiments can be separated into five blocks: stem, encoder, head, decoder, and output. Figure 5.7 shows the connections between each block. The stem block is responsible for reducing the input dimension and adjusting the number of features. It also creates a shortcut connection to the output block before reducing the input dimension.

The encoder block is the most complex of the model. It is responsible for extracting and aggregating features from the volume, enabling the last blocks to generate the segmentation mask. Our encoder block is based on the EfficientNet architecture, explained in Section 5.2.7. We added shortcuts between downsampling layers of the encoder and the upsampling layers on the decoder block to facilitate information flow between different levels of representation, which help to mitigate any spatial information loss by dimensional reduction (Ronneberger *et al.* [2015]).

The head block typically maps the encoder output to the desired output space. In our model, the head block connects the encoder block to the decoder. The decoder block expands the feature map to the original volume size. The output block transforms the feature map into a probability map using the softmax function to create the segmentation probabilities for each voxel.

5.4.5 Efficientnet modifications

The EfficientNet was developed to achieve a balance between performance and resource usage in the image classification problem for the ImageNet dataset. The liver segmentation task shares many characteristics with the image classification problem. However, there are many peculiarities for each

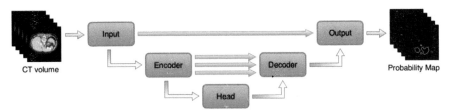

Figure 5.7 Diagram of the network architecture, where grey arrows show the main data flow and blue arrows represent shortcut connections.

problem; with that in mind, we proposed modifications to the EfficientNet architecture to optimize it to our problem.

The first modification was our input method, which receives a sequence of nine slices from the CT volume. Instead of an RGB image as in the original architecture, the input conversion is done by the stem block at the beginning of the network. The network generates a single probability map from the input that corresponds to the segmentation of the middle slice.

The EfficientNet uses inverted residual layers of the MobileNet-V2 as its main building block because, at the time, MobileNet-V2 was the state of the art in terms of accuracy per trainable variables. However, since then, different models have been proposed that reduce resource usage while maintaining the same level of accuracy, including a new version of the MobileNet. We selected the blocks proposed in the ShuffleNet-V2 (Ma *et al.* [2018]) and MobileNet-V3 (Howard *et al.* [2019]) as possible substitutes for the inverted residual layers in our encoder architecture.

A natural way to improve the spatial feature connectivity for volumetric data is to use 3D convolutions. Köpüklü *et al.* [2019] already followed this idea with 3D convolutional versions of MobileNet-V1, MobileNet-V2, ShuffleNet-V1, ShuffleNet-V2, and SqueezeNet, achieving satisfactory results, although their research was focused on motion recognition.

To adapt our model for 3D convolutions, we resized our input to receive eight slices and produce four slices of probability maps as output. Figure 5.8 shows a diagram of the network architecture for 2D and 3D inputs. This was done to guarantee that each probability map had a sufficient amount of neighboring slices to extract information from.

We achieved the desired input and output by performing dimensionality reduction on the z-axis only in the first and third re-scaling and performing upsampling on the z-axis only in the first resizing of the decoder. The shortcut connections are cropped to the desired size on the encoder side to fit the decoder.

5.4.6 Liver postprocessing

We concatenate all the output slices to generate a volume probability map, which allows us to determine the voxel classification through a simple thresholding method. This generates a volumetric mask of our liver prediction that can be used in conjunction with an algorithm to find connected components, which allows us to find the largest connected component and classify it as the liver, thus eliminating most of the small false positives that may occur.

5.4.7 Mask combination

We use the liver mask found in the previous stage to narrow the volume of interest for tumor segmentation. There are many ways to combine the input volume and the liver mask. The chosen approach masks the CT volume with the liver mask, effectively removing any tissue that is not considered liver. This method was chosen due to its simplicity and because it does not increase the input size of the model.

5.4.8 Tumor postprocessing

We use the same methods to create the probability map volume and connected components as in the liver postprocessing step. However, instead of choosing only the largest component, we eliminate all components whose highest probability that a voxel in the component is a tumor is below a threshold.

5.4.9 Network training

Training a network is typically a very time consuming and sensitive process, which can decrease the model performance. With that in mind, we adopted modern techniques to reduce training time, improve model stability, and reduce over-fitting.

Our choice of optimizer was RAdam (Liu *et al.* [2019]), a stochastic adaptive algorithm based on the commonly used Adam (Kingma and Ba [2014]) optimizer. RAdam rectifies its adaptive learning rate to stabilize its variance, which prevents the model from converging to biased/bad local optimal. The RAdam authors based their ideas on the improvement observed in high performance networks that gradually increased their learning rate for a few iterations, before reaching the desired learning rate and resuming their default training schedule. This technique is called linear warmup (Goyal *et al.* [2017]) and aims to stabilize the adaptive learning rate without skewing the network. We chose to use linear warmup in conjunction with RAdam, since the authors Liu *et al.* [2019] who developed RAdam do not explicitly say that the optimizer eliminates the warmup necessity.

For the learning rate scheduler, we use a plateau reduction and look-ahead mechanism (Zhang *et al.* [2019b]). The plateau reduction simply decreases the learning rate by a multiplicative factor when the network has not improved in the last couple of epochs.

The look-ahead algorithm uses two sets of weights: a fast one that updates each iteration and a slow one that updates after k iterations. The slow set is updated with a linear interpolation of its current location and the

Figure 5.8   Diagram expanding the network architecture in the encoder and decoder blocks.

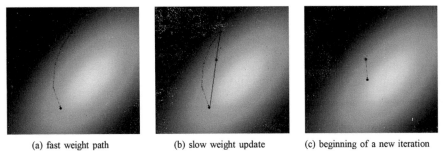

(a) fast weight path   (b) slow weight update   (c) beginning of a new iteration

Figure 5.9   (a) The algorithm after the fast weight updated k times. The black dot shows the fast weight position and the red dot shows the slow weight position. In Figure 5.9(b), we have the new position (in red) for the slow weights. In Figure 5.9(c), we have the first iteration of the fast weight in a new cycle of the algorithm.

position of the fast set of weights. Then, the fast set of weights starts the next iteration from the position of the slow set. Figure 5.9 shows a one-step representation of this process.

We use the weights that achieved the best results in our validation set to avoid over-fitting the training samples. In addition, we use early stopping.

5.4.10 Dataset

The dataset used to evaluate the performance of our method is the LiTS'2017 benchmark, organized with 2017 IEEE International Symposium on Biomedical Imaging (ISBI) and 2017 International Conference on Medical Image Computing and Computer Assisted Intervention (MICCAI).

The data contain liver and tumor labels created manually by trained radiologists and oncologists. The quality of labels correlated to the test set was verified by three experienced radiologists in a blind review (Bilic *et al.* [2019]). The dataset was divided into 131 samples for training and 70 samples for testing. The testing samples were published without the label annotation.

5.5 Experiments

We used two experimental setups: one that is performed locally and the other that is executed on the LiTS challenge server. We opted to carry out local experiments because it allows us to visualize and compare the segmentation results on a case-by-case basis, which in turn helps us to identify the model's shortcomings.

Moreover, the comparison would be very hard to be made with only performance metrics, even with the diversity of metrics reported by the LiTS benchmark. By using a validation set, we also avoid over-fitting our segmentation model to the test set, which could make our model to perform better than a real use-case test would.

5.5.1 Local experiments setup

We divided the LiTS training set into two groups: the training group has 85% of the volumes and the validation group has 15%.

We downsampled the input from 512×512 to 256×256 pixels. Then, we upsampled the predicted masks to the original size for evaluation. We opted to work with smaller image dimensions because it allows for faster experimental iterations. We also limited the experiments to the second stage of our model since several neural networks have already achieved liver segmentation with high Dice scores, while the tumor segmentation still poses a challenging problem.

In our preliminary test using an architecture similar to that proposed by Han [2017], we observed a high variation of the segmentation quality across the volumes. In Figures 5.10 and 5.11, we show the Dice scores per CT volume, where the x-axis is ordered by the average number of voxels per tumor. It is clear that the network performed worse on volumes that presented smaller average tumor sizes.

In order to study the influence of tumor size on the performance of our models, we divided each tumor instance into three groups based on their number of voxels. The value range of each group is shown in Table 5.4, where v stands for the tumor volume.

5.5.1.1 *2D models*

As mentioned in Section 5.4.5, we have three different 2D convolutional layer architectures. For each layer architecture, we experimented with three different sets of hyperparameters: top, default, and bottom.

The default set maintains the same proportions as used in EfficientNet-B0, the top set increases the shallower layers, whereas the bottom set increases the deeper layers (the top and bottom sets have decreased the number of parameters in their deeper and shallower layers to balance the floating point operation (FLOP) count with the default set).

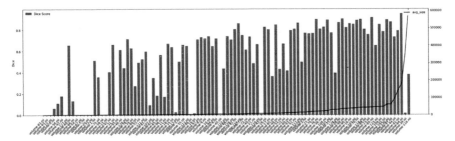

Figure 5.10 Dice score for each individual volume in the training set (ordered by the average size of each tumor). Volumes without any tumors were removed because their Dice values are binary.

Table 5.4 Classification limits used to separate each tumor instance by its size

Group	Volume (voxels)
Small	$v < 1000$
Medium	$1000 \leq v \leq 10,000$
Large	$v > 10,000$

Table 5.5 Number of parameters and FLOPs for each 2D model, calculated for inputs of size $9 \times 256 \times 256$. The values were measured using the library at https://github.com/sovrasov/flops-counter.pytorch

Model	Layer configuration	FLOPs	Parameters
MobileNet-V2	Top	107.14M	4.64M
	Default	123.89M	3.76M
	Bottom	104.80M	6.37M
MobileNet-V3	Top	103.01M	4.47M
	Default	113.72M	3.64M
	Bottom	103.06M	7.80M
ShuffleNet-V2	Top	118.93M	1.44M
	Default	122.31M	1.29M
	Bottom	122.13M	1.47M

We based the choice of our filter size and layer count on the sum of floating point operation. Our target value was the same as the EfficientNet-B0. Table 5.5 shows the number of parameters and FLOPs for all 2D models we tested.

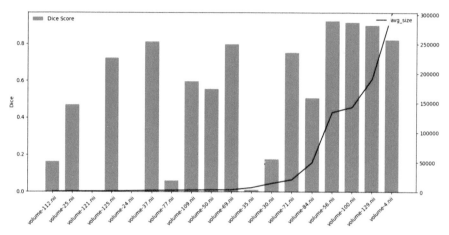

Figure 5.11   Dice score for each individual volume in the validation set (ordered by the average size of each tumor). Volumes without any tumors were removed because their Dice values are binary.

Table 5.6   Number of parameters and FLOPs for each 3D model, calculated for input of size 8 × 256 × 256. These values were obtained using the module from https://github.com/sovrasov/flops-counter.pytorch

Model	Layer configuration	FLOPs	Parameters
MobileNet-V2	Top	64.28M	1.26M
	Default	61.69M	1.59
	Bottom	64.03M	2.93M
MobileNet-V3	Top	67.25M	0.99M
	Default	62.44M	1.95M
	Bottom	64.78M	3.29M
ShuffleNet-V2	Top	68.84M	1.33M
	Default	61.53M	1.59M
	Bottom	56.37M	1.95M
DSTS	Top	66.99M	1.68M
	Default	61.22M	2.52M
	Bottom	55.19M	2.88M

5.5.1.2 *3D models*

In order to reduce memory consumption and computational time introduced by replacing 2D convolutions with 3D convolutions, we reduced the number of channels in each layer by half. Table 5.6 shows the number of parameters and FLOPs for each tested 3D architecture. The number of parameters is much smaller when compared to 2D models, which is expected since the number of channels in the filters was reduced by half.

We tested the same three-layer architectures from the 2D models. We also added the DSTS from Section 5.2.5 and tested three sets of different hyperparameters. They follow the same idea as those used in the 2D model tests, but the default set has its width reduced by half.

5.5.2 Local evaluation

Table 5.7 shows the Dice value for each 2D model separated by tumor size. The results indicate that concentrating the number of parameters in the shallower layers helps in the detection of smaller tumors, while concentrating the parameters in the deeper layers improves large tumor segmentation accuracy.

In Figure 5.12, the left column model (parameters focused on shallow layers) presented problems with larger tumors (top row example) but was able to detect smaller tumors in the second example (middle row). This occurred because more of the shallower layers were able to use the detailed spatial information before the dimensional reduction. The default EfficientNet model (middle column) did not detect as many small tumors as the left column model but performed much better than the deeper model (right column). In the third example (bottom row), both the right and middle models performed better than the left model; however, the right model presented insufficient

Table 5.7 Dice scores for each 2D model separated by tumor sizes. The numbers in bold highlight the best results

Model	Layer configuration	Small	Medium	Large
MobileNet-V2	Top	0.233	0.470	0.615
	Default	0.079	0.182	0.631
	Bottom	0.065	0.293	0.608
MobileNet-V3	Top	0.134	0.405	0.548
	Default	0.140	0.418	0.597
	Bottom	0.104	0.337	0.671
ShuffleNet-V2	Top	0.163	0.385	0.581
	Default	0.058	0.339	0.617
	Bottom	0.143	0.381	0.529

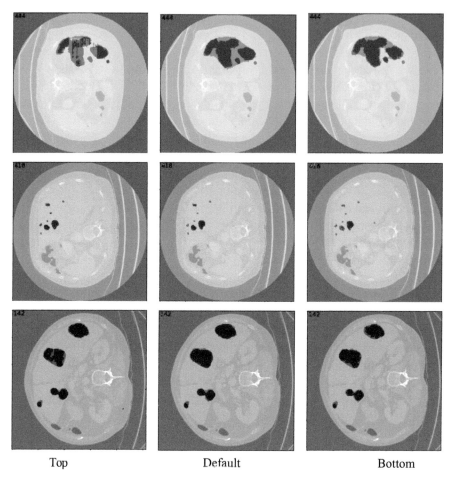

| Top | Default | Bottom |

Figure 5.12   Examples of segmentation for different hyperparameters with the EfficientNet using MobileNet-V2 block. The blue overlay represents the ground-truth segmentation and the red overlay represents the predicted segmentation. The left column represents the top set, the middle column represents the default set, and the right column represents the bottom set. The examples were extracted from volume 4 (slice 444), volume 109 (slice 418), and volume 129 (slice 142). All examples are part of our validation set.

precise spatial information to generate a smoother segmentation, as shown by the straight segmentation lines, which indicate that the model was unable to distinguish those pixels.

Table 5.8 shows the Dice scores for each 3D model separated by tumor size. The results of the 3D models do not clearly indicate how the parameter distribution influences the model performance. According to the results, the best performing models are based on the DSTS layer.

In Figure 5.13, we show segmentation examples for three different architectures: the best 2D architecture, MobileNet-V2-top (2D-MobileNet) on the left column, its 3D counterpart (3D-MobileNet) in the middle column, and the best 3D architecture DSTS-bottom (DSTS) on the right column.

The first example shows a slice where 2D-MobileNet performed a poor segmentation of the large tumor, 3D-MobileNet was unable to detect the large tumor and created an inaccurate segmentation of the smaller one, and the DSTS architecture performed better than both other models, with a better segmentation of the larger tumor and an accurate segmentation of the smaller one. However, the results were still worse than the best 2D architecture shown in Figure 5.12. The jittering artifact can be explained by an underfitting of the architecture, caused by the reduced number of parameters.

In the second example, we have a group of small tumors. All architectures performed well in this example. The 2D-MobileNet had the best results, followed by DSTS, and, finally, 3D-MobileNet. Moreover, both 3D architectures detected some tumors better than the 2D architectures, which indicates that some tumors are easier to segment with 3D architectures.

In the third example, both MobileNet architectures generated an inferior segmentation in the large tumor compared to the DSTS architecture, and the 2D-MobileNet produced a better segmentation for the small tumor at the bottom.

Table 5.8 Dice scores of each 3D model separated by tumor sizes. The numbers in bold highlight the best results

Model	Layer configuration	Small	Medium	Large
MobileNet-V2	Top	0.195	0.455	0.646
	Default	0.125	0.359	0.577
	Bottom	0.104	0.388	0.624
MobileNet-V3	Top	0.196	0.415	0.646
	Default	0.181	0.392	0.619
	Bottom	0.200	0.472	0.631
ShuffleNet-V2	Top	0.151	0.405	0.535
	Default	0.243	0.454	0.611
	Bottom	0.122	0.406	0.582
DSTS	Top	0.192	0.483	0.631
	Default	0.151	0.447	0.617
	Bottom	0.289	0.572	0.668

The fourth example shows a tumor that 2D-MobileNet failed to detect, but both 3D architectures segmented
well. The last example shows a small false positive from 2D-MobileNet that is not present in the 3D architectures.

5.5.3 Lits challenge evaluation

We have submitted three predictions to the LiTS challenge: the first was based on the MobileNet-V2 with 2D convolutions, the second was based on the MobileNet-V2 with 3D convolutions, and the last used the DSTS layers. We trained the models using the same splits as the local evaluation.

The evaluation of the LiTS challenge is done on their servers. The submission is composed of the mask for each volume, where the value 0 represents background, 1 represents liver, and 2 represents the tumor. The number of submissions is limited to 3 per day to avoid model bias. The metadata from the submitted Neuroimaging Informatics Technology Initiative (NIfTI) files must be appropriate to the corresponding CT volume. Table 5.9 lists the LiTS challenge entries alongside their segmentation evaluation metrics and parameter count.

Table 5.9   Segmentation metrics of the best performing published works on top and our model results separated by rule at the bottom. The ↑ indicates that results with higher values are better; ↓ indicates that results with lower values are better; and → 0 indicates that results closer to 0 are better. Our models are at the bottom separated by a rule.

Model	Dice per Case (↑)	Global Dice (↑)	VOE (↓)	RVD (→0)	ASSD (↓)	Parameters (↓)
Volumetric attention Wang *et al.* [2019b]	0.741	0.813	0.389	−0.177	1.1224	25M+
LW HCN Zhang *et al.* [2019a]	0.730	0.820	NA	NA	NA	3.6M
H-DenseUNet Li *et al.* [2018]	0.722	0.824	0.366	4.272	1.102	80M
DeepX Yuan [2017]	0.657	0.820	0.378	0.288	1.151	30M
3D-UNet Ç içek *et al.* [2016]	0.55	NA	NA	NA	NA	19M
X.Han Han [2017]	NA	0.670	0.450	0.040	6.660	24M
DSTS-3D (ours)	0.548	0.768	0.400	−0.120	1.187	2.88M
2D MobileNet-V2 (ours)	0.536	0.678	0.429	−0.162	1.4232	4.64M
3D MobileNet-V2 (ours)	0.524	0.746	0.393	−0.041	1.1298	1.26M

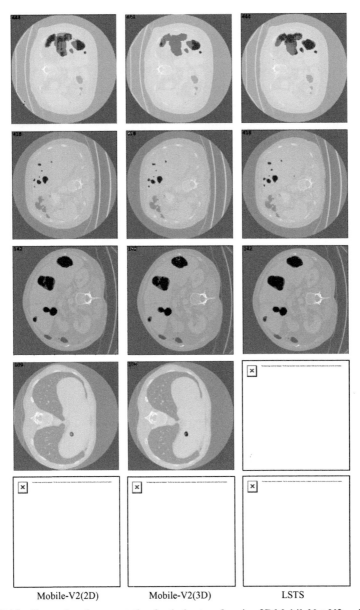

Mobile-V2(2D)   Mobile-V2(3D)   LSTS

Figure 5.13   Examples of segmentation for the best performing 2D MobileNet-V2 architecture (left), its 3D counter-part (middle), and the best performing 3D architecture (right). The blue overlay represents the ground-truth segmentation and the red overlay represents the predicted segmentation. The three first examples are the same as in Figure 5.12. The fourth example was extracted from volume 35 (slice 109) and the last example was extracted from volume 69 (slice 203), both from the validation set.

Our method did not achieve the performance necessary to compete with more specialized liver tumor seg mentation methods. However, unlike most approaches, our method focused on limited resource environments, instead of maximizing performance.

Our best performing method achieved performance similar to 3D-UNet with 5 times less parameters. It also achieved proper results in the VOE and ASSD metrics, but its weakest metric was given by the Dice per case. If we compare the 2D MobileNet-V2 with 3D MobileNet-V2, we notice a drop in Dice per case but an increase in ASSD, VOE, and Dice global, which indicates that the 3D model created better overall segmentation, but the 2D model detected tumors correctly in more volumes.

The limited number of FLOPS that we imposed on our 3D models reduced its ability to detect smaller tumors. The only best performing solution with comparable number of parameters was the LW HCN model, but it uses images four times larger than ours to generate its prediction.

Conclusion

In this work, we studied and analyzed the viability of efficient networks as encoder backbones for the liver tumor segmentation problem. We started from a baseline network to investigate its shortcomings and create a better understanding of the challenges posed by our problem.

The segmentation quality was correlated with the tumor size present in the volume. Large tumors presented a high segmentation quality and detection rate, while small tumors were poorly detected or overestimated. Based on our observations, we proposed a method to evaluate the segmentation quality that distinguishes tumors by their size, which allows a better understanding of how the model works under different circumstances.

We created two models, one based on 2D convolutions and the other based on 3D convolutions. Our 2D model is end-to-end segmentation based on the U-Net architecture with an adaptation of the EfficientNet-B0 as an encoder. We tested two additional efficient layer architectures for the inverted residual layer (EfficientNet base layer), with three different layer distributions to determine how optimized the original EfficientNet architecture was for the tumor liver segmentation problem. Our results demonstrated that focusing on layers before the dimensional reduction improved the detection of small tumors, with some degradation of the large tumor segmentation.

Our 3D-convolution model followed the same architecture as the 2D model. However, the input and output were adapted to fit the 3D convolutional filters. We repeated the experiments with the same layers as the 2D model and added another layer architecture, referred to as DSTS layer, described in Section 5.2.5. The DSTS model had the best performance, improving in all tumor sizes compared to any 3D model, but it still had issues that were not present in some 2D models.

Both 2D and 3D models were able to create segmentation predictions in less than 1 second per slice, on a machine without using a GPU. However, 3D models were almost ten times slower than their 2D counterparts, even with a lower number of parameters and FLOP count. We attribute this slowdown to the number of dependencies created by the 3D convolutions. Although the 3D models had some advantages over the 2D models, they were not enough to justify the reduced performance.

References

K. S. Albain, R. S. Swann, V. W. Rusch, A. T. Turrisi III, F. A. Shepherd, C. Smith, Y. Chen, R. B. Livingston, R. H. Feins, and D. R. Gandara. Radiotherapy plus Chemotherapy with or Without Surgical Resection for Stage III Non-Small-Cell Lung Cancer: a Phase III Randomised Controlled Trial. The Lancet, 374(9687): 379–386, 2009.

P. Bilic, P. F. Christ, E. Vorontsov, G. Chlebus, H. Chen, Q. Dou, C.-W. Fu, X. Han, P.-A. Heng, and J. Hesser. The Liver Tumor Segmentation Benchmark (LiTS). arXiv preprint arXiv:1901.04056, 2019.

H. Chen, Z. Qin, Y. Ding, L. Tian, and Z. Qin. Brain Tumor Segmentation with Deep Convolutional Symmetric Neural Network. Neurocomputing, 392:305–313, 2020. L.-C. Chen, G. Papandreou, I. Kokkinos, K. Murphy, and A. L. Yuille. Deeplab: Semantic Image Segmentation with Deep Convolutional Nets, Atrous Convolution, and Fully Connected CRFS. IEEE Transactions on Pattern Analysis and Machine Intelligence, 40(4):834–848, 2017a.

L.-C. Chen, G. Papandreou, F. Schroff, and H. Adam. Rethinking Atrous Convolution for Semantic Image Segmentation. arXiv preprint arXiv:1706.05587, 2017b.

S. Chen, C. Ding, and M. Liu. Dual-Force Convolutional Neural Networks for Accurate Brain Tumor Segmen-tation. Pattern Recognition, 88:90–100, 2019.

Ö. Ç i̧ cek, A. Abdulkadir, S. S. Lienkamp, T. Brox, and O. Ronneberger. 3D U-Net: Learning Dense Volumet-ric Segmentation from Sparse Annotation. In International Conference on Medical Image Computing and Computer-Assisted Intervention, pages 424–432. Springer, 2016.

L. R. Dice. Measures of the Amount of Ecologic Association Between Species. Ecology, 26(3):297–302, 1945.

R. C. Gonzalez, R. E. Woods, and S. L. Eddins. Digital Image Processing Using MATLAB. Pearson Education India, 2004.

P. Goyal, P. Doll´ar, R. Girshick, P. Noordhuis, L. Wesolowski, A. Kyrola, A. Tulloch, Y. Jia, and K. He. Accurate, Large Minibatch SGD: Training Imagenet in 1 hour. arXiv preprint arXiv:1706.02677, 2017.

X. Han. Automatic Liver Lesion Segmentation Using a Deep Convolutional Neural Network Method. arXiv preprint arXiv:1704.07239, 2017.

K. He, X. Zhang, S. Ren, and J. Sun. Deep Residual Learning for Image Recognition. In IEEE Conference on Computer Vision and Pattern Recognition, pages 770–778, 2016.

K. He, G. Gkioxari, P. Doll´ar, and R. Girshick. Mask R-CNN. In IEEE International Conference on Computer Vision, pages 2961–2969, 2017.

G. N. Hounsfield. Computerized Transverse Axial Scanning (Tomography): Part 1. Description of System. The British Journal of Radiology, 46(552):1016–1022, 1973.

A. Howard, M. Sandler, G. Chu, L.-C. Chen, B. Chen, M. Tan, W. Wang, Y. Zhu, R. Pang, and V. Vasudevan. Searching for Mobilenetv3. In IEEE International Conference on Computer Vision, pages 1314–1324, 2019.

A. G. Howard, M. Zhu, B. Chen, D. Kalenichenko, W. Wang, T. Weyand, M. Andreetto, and H. Adam. Mobilenets: Efficient Convolutional Neural Networks for Mobile Vision Applications. arXiv preprint arXiv:1704.04861, 2017.

J. Hu, L. Shen, and G. Sun. Squeeze-and-Excitation Networks. In IEEE Conference on Computer Vision and Pattern Recognition, pages 7132–7141, 2018.

P. Jaccard. Distribution de la Flore Alpine Dans le Bassin des Dranses et Dans Quelques R´egions Voisines. Bull Soc Vaudoise Sci Nat, 37:241–272, 1901.

D. P. Kingma and J. Ba. Adam: A Method for Stochastic Optimization. arXiv preprint arXiv:1412.6980, 2014.

O. Köpüklü, N. Kose, A. Gunduz, and G. Rigoll. Resource Efficient 3D Convolutional Neural Networks. In IEEE/CVF International Conference on Computer Vision Workshop, pages 1910–1919. IEEE, 2019.

D. C. Le, K. Chinnasarn, J. Chansangrat, N. Keeratibharat, and P. Horkaew. Sema-Automatic Liver Segmen-tation based on Probabilistic Models and Anatomical Constraints. Scientific Reports, 11(1):1–19, 2021.

X. Li, Z. Liu, P. Luo, C. Change Loy, and X. Tang. Not All Pixels are Equal: Difficulty-Aware Semantic Segmentation Via Deep Layer Cascade. In IEEE Conference on Computer Vision and Pattern Recognition, pages 3193–3202, 2017.

X. Li, H. Chen, X. Qi, Q. Dou, C.-W. Fu, and P.-A. Heng. H-DenseUNet: Hybrid Densely Connected UNet for Liver and Tumor Segmentation from CT Volumes. IEEE Transactions on Medical Imaging, 37(12):2663–2674, 2018.

L. Liu, H. Jiang, P. He, W. Chen, X. Liu, J. Gao, and J. Han. On the Variance of the Adaptive Learning Rate and Beyond. arXiv preprint arXiv:1908.03265, 2019.

T. Livraghi, A. Giorgio, G. Marin, A. Salmi, I. De Sio, L. Bolondi, M. Pompili, F. Brunello, S. Lazzaroni, and G. Torzilli. Hepatocellular Carcinoma and Cirrhosis in 746 Patients: Long-Term Results of Percutaneous Ethanol Injection. Radiology, 197(1):101–108, 1995.

N. Ma, X. Zhang, H.-T. Zheng, and J. Sun. Shu enet v2: Practical Guidelines for Efficient CNN Architecture Design. In European Conference on Computer Vision, pages 116–131, 2018.

L. Meng, Q. Zhang, and S. Bu. Two-Stage Liver and Tumor Segmentation Algorithm Based on Convolutional Neural Network. Diagnostics, 11(10):1806, 2021.

F. Milletari, N. Navab, and S.-A. Ahmadi. V-net: Fully Convolutional Neural Networks for Volumetric Medical Image Segmentation. In Fourth International Conference on 3D Vision, pages 565–571. IEEE, 2016.

O. Ronneberger, P. Fischer, and T. Brox. U-Net: Convolutional Networks for Biomedical Image Segmentation. In International Conference on Medical Image Computing and Computer-Assisted Intervention, pages 234–241. Springer, 2015.

S. Rossi, M. Di Stasi, E. Buscarini, P. Quaretti, F. Garbagnati, L. Squassante, C. Paties, D. Silverman, and L. Buscarini. Percutaneous RF Interstitial Thermal Ablation in the Treatment of Hepatic Cancer. AJR. American Journal of Roentgenology, 167(3):759–768, 1996.

H. Sung, J. Ferlay, R. L. Siegel, M. Laversanne, I. Soerjomataram, A. Jemal, and F. Bray. Global cancer statistics 2020: GLOBOCAN estimates of incidence and mortality worldwide for 36 cancers in 185 countries. CA: A Cancer Journal for Clinicians, 0:1–41, 2020.

M. Tan and Q. V. Le. Efficientnet: Rethinking Model Scaling for Convolutional Neural Networks. arXiv preprint arXiv:1905.11946, 2019.

P. Tang, Q. Liang, X. Yan, S. Xiang, W. Sun, D. Zhang, and G. Coppola. Efficient Skin Lesion Segmentation Using Separable-Unet With Stochastic Weight Averaging. Computer Methods and Programs in Biomedicine, 178:289–301, 2019.

K. D. Toennies. Guide to Medical Image Analysis. Springer, 2017.

P. Tschandl, C. Sinz, and H. Kittler. Domain-Specific Classification-Pretrained Fully Convolutional Network Encoders for Skin Lesion Segmentation. Computers in Biology and Medicine, 104:111–116, 2019.

G. Wang, W. Li, T. Vercauteren, and S. Ourselin. Automatic Brain Tumor Segmentation Based on Cascaded Convolutional Neural Networks with Uncertainty Estimation. Frontiers in Computational Neuroscience, 13:56, 2019a.

X. Wang, R. Girshick, A. Gupta, and K. He. Non-Local Neural Networks. In IEEE Conference on Computer Vision and Pattern Recognition, pages 7794–7803, 2018.

X. Wang, S. Han, Y. Chen, D. Gao, and N. Vasconcelos. Volumetric Attention for 3D Medical Image Seg-mentation and Detection. In International Conference on Medical Image Computing and Computer-Assisted Intervention, pages 175–184. Springer, 2019b.

F. Xie, J. Yang, J. Liu, Z. Jiang, Y. Zheng, and Y. Wang. Skin Lesion Segmentation Using High-Resolution Convolutional Neural Network. Computer Methods and Programs in Biomedicine, 186:105241, 2020.

Y. Yuan. Hierarchical Convolutional-Deconvolutional Neural Networks for Automatic Liver and Tumor Seg-mentation. arXiv preprint arXiv:1710.04540, 2017.

J. Zhang, Y. Xie, P. Zhang, H. Chen, Y. Xia, and C. Shen. Light-Weight Hybrid Convolutional Network for Liver Tumour Segmentation. In 28th International Joint Conference on Artificial Intelligence, Macao, China, pages 10–16, 2019a.

M. Zhang, J. Lucas, J. Ba, and G. E. Hinton. Lookahead Optimizer: k Steps Forward, 1 Step Back. In Advances in Neural Information Processing Systems, pages 9593–9604, 2019b.

6

Advancements in Deep Learning Techniques for Analyzing Electronic Medical Records

K. Manimala[*1], **E. Fantin Irudaya Raj**[2], **and S. Jagatheswari**[3]

[*1]Professor, Dr. Sivanthi Aditanar College of Engineering, India;
*Email: smonimala@gmail.com
[2]Assistant Professor, Dr. Sivanthi Aditanar College of Engineering, India;
Email: fantinraj@gmail.com
[3]Assistant Professor (Sr), Vellore Institute of Technology, India;
Email: jagatheswari.s@vit.ac.in
*Corresponding author: K. Manimala

Abstract

In recent years, there has been a huge amount of data collected in hospitals in terms of electronic medical records (EMR), continuity of care documents (CCD), electronic healthcare records (EHR), and many more. These records contain information about patients such as their location, symptoms, diagnosed disease, blood test results, scan images and reports, recommended medications, and doctor's notes. The aforementioned details can assist physicians and care providers in providing better healthcare solutions for the end-user. On the other hand, the amount of data available has increased tremendously, necessitating the use of analytical, clinical, and business intelligence tools to transform it into useful information. Researchers are attempting to integrate the latest breakthroughs in deep learning technologies to extract clinical information from data traditionally used for local billing and archiving purposes, which is advantageous to people. Because these records were created for the sole purpose of local administration and hospital management, they were unstructured, necessitating deep learning research, which has been shown effective in a number of other domains because of its efficacy in capturing data dependencies. Clinical risk prediction based on such big data analysis of healthcare records will undoubtedly aid in the prediction of undesirable occurrences such as cardiac arrest, lowering patient mortality rates. Several researchers are working on this topic in order to provide better patient care and gain more insights into how to enhance healthcare in the future. The most difficult aspect of dealing with these records is the disparity in how patient data is kept, such as weight as numbers, admission data as dates and times, categorical values for diagnosis, natural language for doctor's remarks, and so on. Other issues, such as sparsity, missing data, and large data dimensionality, require specific research attention. As a result, more research is needed regarding mining these massive amounts of data; at present, research has shortcomings in model design, deep learning technique deployment, and the lack of globally acknowledged assessment standards that are highly needed. The present work examines the current state of the art, existing deep learning approaches used, the research gap, and recommendations for improved deep-learning deployment in EHR research.

6.1 Introduction

Energy almost all the hospitals of the entire world have adopted electronic healthcare record (EHR) systems either at the basic level or at a high level to support clinical decision support systems. Several research works are reported for the analysis of these EHR records which used varied techniques including machine learning, regression, and support vector machines to the recent popular deep learning techniques [1]. Hence, there is a necessity to review these data analysis techniques and draw inferences to analyze the progress of clinical applications due to the advancements of these learning techniques. A specific summary on the application of deep learning techniques for EHR records is necessary as there are several reports of the usage of deep learning in the broad area of medical fields like CT, MRI, and ECG image analysis. Especially the total unstructured arrangement of data and the variety of sources through which these data are collected necessitated the entry of deep learning techniques into this area.

Trusted systems should be designed to analyze such heterogeneous data that will learn the huge clinical data and learn the correlation between diseases and their associated symptoms. In addition to that, the causes for the diseases, the number of people usually affected by a specific disease, the regions usually affected, the risk factors of such diseases, etc., should be identified properly to reduce the impact of such disease in the future. The confidentiality of the patient's data should be maintained while doing such rigorous analysis. Several conventional machine learning methods like principal component analysis (PCA) [2], [3], artificial neural networks (ANN), convolutional neural networks (CNN), Bayesian network, probabilistic neural network, support vector machine (SVM) [4], [5], decision trees [6], etc., are employed for the analysis of this clinical big data. These methods have produced successful results in several sub-areas of hospital data but the highly complicated and heterogeneous EHR data requested the need for advanced deep learning techniques which has achieved success in several other contemporary areas. Deep learning ensures accuracy in prediction and knowledge dissemination in several areas, and, hence, there is no surprise in the usage of these techniques for getting precision in clinical analysis.

6.2 Overview of EHR

Hospitals are competing with each other to provide better care for patients by maintaining their history of healthcare problems and better coordination among health experts through electronic records. Such records really benefitted both the patients and hospitals by reducing errors and improving the efficiency of patients care by identifying the earlier health issues, the medicines being taken, and the general health condition within a shorter duration. EHRs are generally categorized into two, namely the basic EHR which lacks doctor's notes and the advanced comprehensive EHRs which contains doctor's notes. It was designed with an aim of administrative convenience within hospitals, and, hence, there exist several categories of coding for entering the health data of patients. There are certain codes like International Statistical Classification of Diseases (ICD) and related health problems, Current Procedure Terminology (CPT), and Logical Observation Identifiers Names and Codes (LOING) [1]. The above said codes are not fixed, but it varies with respect to different hospitals and different parts of the world. Deep learning techniques will work effectively only when the codes are properly represented in EHR. Due to the huge amount of data and the different varieties of codes, the coordination and analysis becomes a tough task that necessitated the introduction of deep learning technologies into this area.

6.2.1 Characteristics of EHR

Vast amount of data is collected as EHR in hospitals all over the world and it is reported that such data has exceeded 25 exabytes [7], [8]. Thanks to Internet of Things (IoT), several wearable sensors have been developed to monitor the health of patients automatically which contributes to the huge surge in such data. Real-time data is also generated from such sensors which require immediate analysis and fast action. On the one side there is a growth in quantity of data, and on the other side, there is a growth in the number of sources from which the data is collected. As the number of sources increases, the complexity of analysis increases as there is not any common way to describe the codes of diseases and the format of clinical notes. Thus, it can be concluded clearly that EHR data has all the physical characteristics of big data in terms of velocity, veracity, variety, and volume, and, hence, the big data analysis should be extended to it.

6.2.2 Categories of EHR data

Huge data are collected during the patient's health check-up or treatment process. This information is represented and stored in different formats by the hospitals. Medical imaging, diseases coding, and clinical notes are the basic categories of data. There are several imaging techniques for medical diagnosis namely the optical coherence tomography (OCT), magnetic resonance imaging (MRI), computed tomography (CT), X-rays, and signal-based data such as EEG, ECG, etc. MRI images are usually used to identify the disorder in the brain, heart, bones, and joints. CT and X-rays are normally used to detect abnormalities of the abdomen and chest. However, the difficulty of analysis of clinical data lies in the availability of such data compared to the other big data like the social website data. Hence, it is not possible to provide the required amount of data for the training phase. Another problem associated with this data is the format in which these data can be presented to the layers of the neural networks.

6.2.3 Doctor's notes

This is in the form of text like discharge summary, scan reports, death certificates, etc. [9]–[18]. An in-depth study of discharge summary is usually done as it contains more information like diagnosis, the conclusion of lab test studies, drugs taken during admission, medicines to be taken, and the follow-up action is required.

6.3 Machine Learning and Deep Learning in EHR

Initially, researchers used machine learning techniques for the knowledge discovery process of EHR record. The process of making the machine intelligent by making them learn from existing data is called machine learning. Supervised and unsupervised learning are the two common types of machine learning. The process of mapping input with the already known targets is termed supervised learning and the method of grouping related data without providing a target is termed unsupervised learning. Regression and classification are grouped under machine learning and the clustering process is categorized under unsupervised learning. Significant features need to be extracted from data that forms the input of the machine learning algorithm. Extracting such significant features is one of the crucial issues of the machine learning process and should be addressed properly. In some cases, these features are extracted manually, but that is not possible with EHR data due to its voluminous and heterogeneous nature. Hence, the machine learning or deep learning approaches of these records require automatic feature extraction. These features are given as input to the input layer of the neural network used as a machine learning technique. The commonly used architectures for machine learning and deep learning of EHR data are summarized below. Figure 6.1 shows the basic feed-forward network used for machine learning.

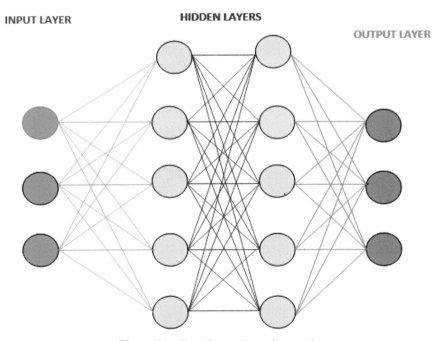

Figure 6.1 Feed-forward neural network.

6.3.1 Multilayer perceptron (MLP) network

The multilayer perceptron is one of the commonly used networks for learning patterns and behavior of EHR data. It usually consists of several hidden layers; the number of hidden layers being decided by the complexity of the problem. It is a general feed-forward network where each and every neuron is connected to all neurons in the next higher layer. The flow of information is only in one direction only and there is no feedback connection. The neuron in each layer sums up the multiplied values of input and connection weights from the previous layer and applies the activation function to produce the final output. The activation functions usually used are the threshold, linear, and sigmoid activation functions. The networks learn the weights of connection between neurons by finding the relation between the input and the output.

The hidden layer weights are updated as shown in eqn (6.1), where d represents the number of units in the previous layer, x_j is the input from the previous layer node j, w_{ij} and b_{ij} represent the weight and biases, and σ represents the activation function [1]. Similarly, the output unit updates its connection weights and the final output is calculated.

$$h_i = \sigma\left(\sum_{j=1}^{d} x_j w_{ij} + b_{ij}\right). \tag{6.1}$$

Figure 6.2 shows the multilayer perceptron network.

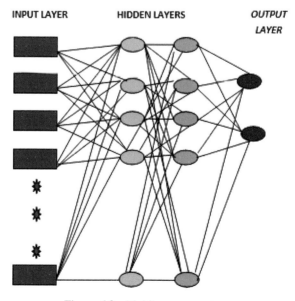

Figure 6.2   Multilayer perceptron.

6.3.2 Convolutional neural network (CNN)

CNN is a tool that has gained good application in fields like image processing especially medical images analysis, computer vision, data mining, etc. The significant attribute of CNN is its ability to identify strong connectivity among neighboring data instead of viewing the raw data as a whole. CNN is a learning mechanism that extracts strong lower-level features from proceeding layers and passes them on to the next higher layer that comprises convolutional filters. The extracted features are summarized in the pooling layer. It is the first and foremost tool commonly used in clinical applications. U-Net [19] and ResNet [20] are the commonly used architectures for medical image analysis that can represent full image with less number of layers and they have good convergence. Dense CNNs [21] have reduced the gradient problem and also the parameters involved. CNNs are proved to be efficient in predicting disease codes from doctor's notes. The basic architecture of CNN is shown in Figure 6.3.

The one-dimensional convolution [1] is given by

$$C_{1d} = \sum_{a=-\infty}^{\infty} x(a)\, w(t-a). \tag{6.2}$$

The two-dimensional convolution [1] is

$$C_{2d} = \sum_m \sum_n X(m,n)\, K(i-m, j-n) \tag{6.3}$$

where x is the input signal, w is the weighting function, X is a 2D grid, and K is a kernel.

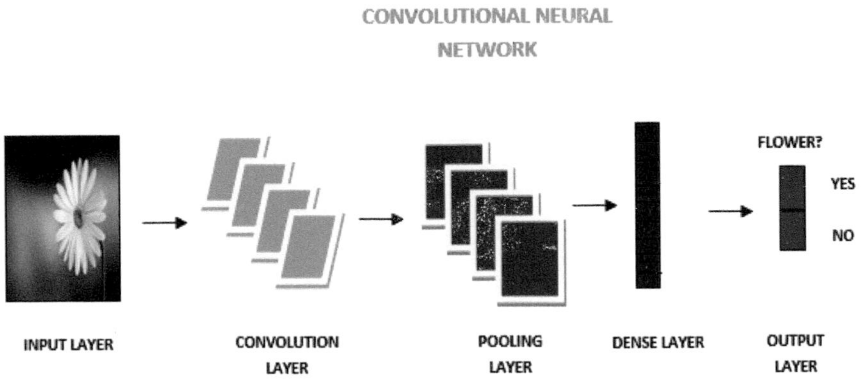

CONVOLUTIONAL NEURAL
NETWORK

| INPUT LAYER | CONVOLUTION LAYER | POOLING LAYER | DENSE LAYER | OUTPUT LAYER |

Figure 6.3   Convolutional neural networks.

6.3.3 Recurrent neural network (RNN)

It is a feed-forward neural network used for deep learning. The strong connections of this type of neural network make it look like a directed graph. Gate recurrent unit is one version of RNN variant which is referred to as long short-term memory (LSTM). Researchers have applied this network for diagnosis of diseases [24], [25] and other temporary medical advice. Gated RNNs are efficient in modeling long-term dependencies [26]. In addition to the medical-related problems, this network has been proved to be successful in other fields like pattern recognition and several image processing applications [27]. RNN is the best choice compared to CNN when the data is arranged in proper order. The network is designed in a way to deal with temporal dependencies [1], and, hence, it is commonly preferred for natural language processing and time series data. The RNNs update its hidden state based on the current input as well as the previous hidden state. Thus, it is

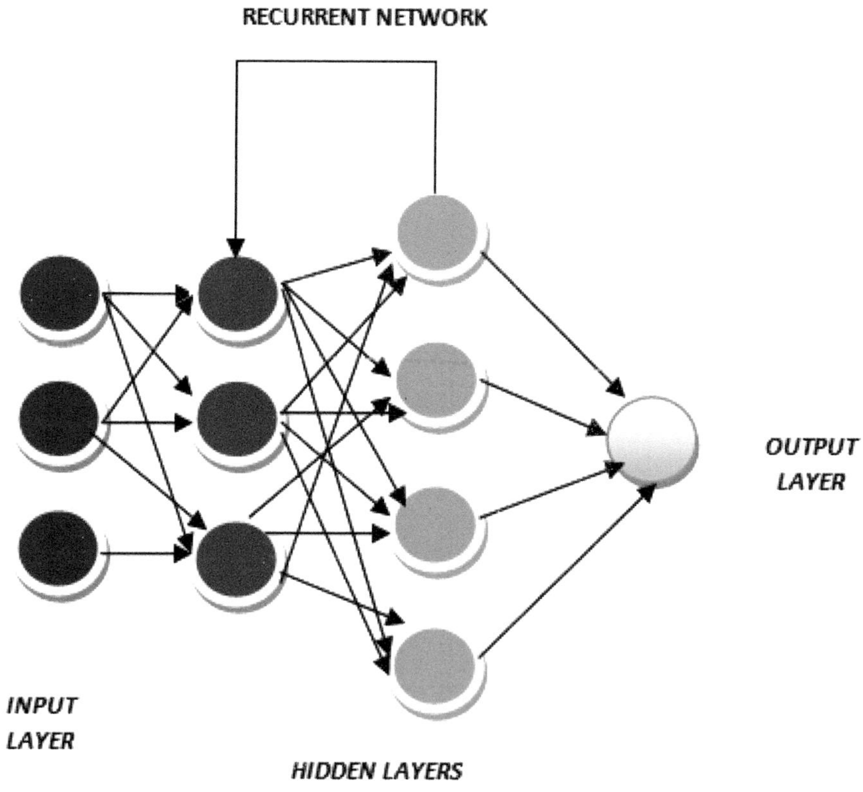

Figure 6.4 Recurrent neural networks.

clear that the final state contains the details of all the previous stages. RNN is also used for extracting features from clinical data. Figure 6.4 represents the architecture of RNN.

6.3.4 Restricted boltzmann machine

It is an unsupervised stochastic learning technique which recognizes the probability distribution of input data. This network finds application in the field of classification tasks of medical images like X-rays [28] and CT images [29]. RBMs are generalized models that interpret the basic process of data generation. In general, a Boltzmann machine is fully connected, whereas in RBN, the connections do not exist between the units of hidden layer or between the visible units. Gibb's sampling is the process used to train RBM. Deep belief network (DBN) is the stacked representation of RBN to form a supervised learning network.

The energy function of RBN [26] is given by

$$E(v, h) = -b^T v - C^T h - W v^T h \qquad (6.4)$$

where v represents visible units and h the hidden units. Figure 6.5 shows the visible and hidden units of RBM.

6.3.5 Autoencoders

Autoencoders (AEs) find application in several pre-training processes of supervised tasks when there is little training data but is an epitomizing unsupervised deep learning technique. AE projects the input data into

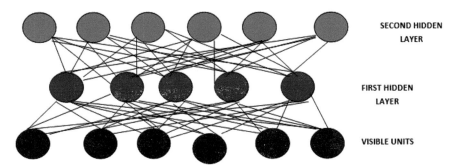

Figure 6.5   Restricted Boltzmann machine.

lower-dimensional space and this is termed as encoding. The representation of the encoding and reconstruction mechanism [1] is given in eqn (6.5) and (6.6).

$$z = \sigma(Wx + b) \tag{6.5}$$

$$\tilde{x} = \sigma\left(W'z + b'\right) \tag{6.6}$$

where x is the input, $\tilde{x}$ is the approximate representation of input, W is the encoding weight, W' is the decoding weight, and z is the encoded representation. The network is trained to reduce the error $\|x - \tilde{x}\|$, thereby making the encoded representation accurate. AE is treated as a dimensionality reduction technique similar to principal component analysis (PCA) as it converts input data into discriminant aspects and stores them as output. It has an edge over PCA and similar techniques in solving complex problems because of the nonlinear transformations that occur in the activation functions of hidden layers. Stacking-based greedy training method is used to design deep AE network. Figure 6.6 shows the architecture of autoencoder.

6.4 Deep Learning Analysis of EHR

The commonly used deep learning architectures are shown in Figure 6.7. The networks used in literature for electronic healthcare analysis are artificial neural network (ANN), convolutional neural network (CNN), deep neural

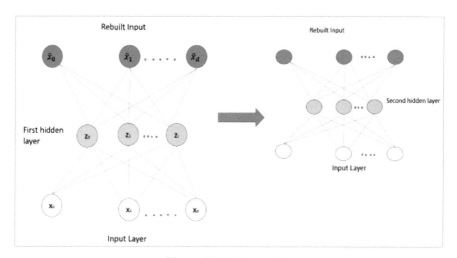

Figure 6.6 Autoencoder.

network (DNN), deep belief network (DBN), stacked autoencoder (SAE), recurrent neural network (RNN), etc. The tasks for which these networks were used [8] are shown in Table 6.1.

Table 6.1   Example of medical codes

Category	Technique	Medical task
Classification	Deep CNN	Old age-related problems detection
		Retinopathy identification
		Macular worsening and diabetic retinopathy identification
Image detection	Fast CNN	Detection of hemorrhage
Risk factor detection	DNN	Cardiovascular risk factors
Image segmentation	CNN	Optic cup and optic disc categorization
Classification	CNN	Disease diagnosis position
	DNN	Alzheimer's disease detection
	CNN and RNN	Alzheimer's disease detection
Classification	CNN	Skin cancer categorization
Classification	DBN	Health control category identification
	SAE	Schizophrenia and Huntington's disease classification
	CNN	Alzheimer's disease and health control classification
	ANN	Alzheimer's disease and health control classification
Image detection	CNN	Vertebral discs analysis and lesion hot spots detection
		Left ventricle sliced detection
		Vertebrae labeling
	RNN	Identification of end-systole frames
Image segmentation	DBN	Ventricle segmentation
	CNN	Prostate lesions segmentation
		Knee cartilage segmentation
Image detection and segmentation	CNN	Vertebrae segmentation

Table 6.1   Continued

Category	Technique	Medical task
Survival prediction	DNN	Survival prediction
Image segmentation	CNN	Segmentation of organs
		Pancreas segmentation
		Liver segmentation
		Kidney segmentation
Image detection	CNN	Colon polyp detection
Classification	CNN	Lung texture classification
		Interstitial pattern classification
	DBN and RBM	Classification of lung texture
Mortality prediction	CNN	Old age people mortality prediction
Classification	CNN	Frontal/lateral classification
		Abnormalities detection
		Pathology detection
		Discriminate malignant masses from cysts
		Tuberculosis detection
		Detection of cardiovascular disease
		Tissue classification
	SAE	Breast density classification
Image detection	CNN	Mass detection

6.4.1 Extraction of information

The toughest part of EHR for extracting information is the doctor's notes in the form of the discharge summary, diagnosis, etc., which is generally documented for further follow-up action. The record of each patient contains information like admission date, address and contact details, X-rays, scan reports, and finally their discharge reports. Due to the unstructured and heterogeneous nature of such data, the process of extracting useful information is highly complex. Hence, manual intervention is necessary to support information extraction. Many recent works were reported for extracting significant attributes from clinical data based on deep learning techniques.

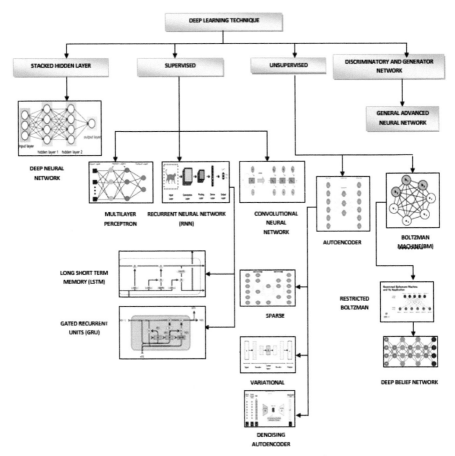

Figure 6.7   Commonly used deep learning architectures.

Such information extraction step consists of the following chores, namely the concept extraction, time event extraction, correlation identification, and, finally, the acronym expansion. Figure 6.8 shows the steps of EHR information extraction.

6.4.1.1 *Concept extraction*

The main task is to extract the basic concept from EHR like the diagnosis of disease, the drugs given, and the follow-up actions. Initially, natural language processing was implemented by the researchers for this purpose, and the results need further improvement due to the complexity of doctor's notes. The researchers [32], [33] used the technique of assigning certain tags to

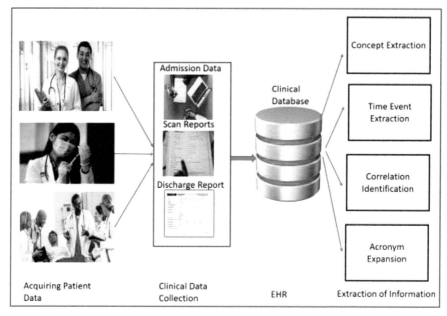

Figure 6.8 EHR information extraction.

each and every word of the note. The tags were categorized based on the type of disease and the drugs suggested. Deep architectures based on RNNs and LSTMs were reported for information extraction.

6.4.1.2 *Time event extraction*
Each and every patient data should be associated with the time of admission, duration of treatment, discharge date, and the time of follow-up actions. RNN network [34] was used to extract the time event from EHR records. Deep dive application is used to identify the structured relationship and forecasting in the future [35].

6.4.1.3 *Correlation extraction*
The relationships between the medical parameters are to be identified to finally establish the cause–effect theory. It is necessary to clearly identify the conditions such as increase/decrease in parameter A results in disease X or test B results indicate disease Y. Such correlations need to be ascertained clearly for extracting features to be given as input to the classifier [1].

6.4.1.4 *Acronym expansion*
There exist nearly 1.97 lakhs of acronyms that need expansion for information extraction. This is actually a tough task as each acronym may have more than

10 possible expansions. Word embedding is the commonly used approach to tackle this challenging task [36].

6.4.2 EHR representation

EHR contains several codes that map to the respective diseases of patients. The codes are generally used for administrative purposes within the hospital, but the researchers find its use for extracting important information which could be useful for the benefit of mankind. Manual mapping of basic symptoms with their respective codes is really a tough task as several symptoms map to same codes and nearly similar symptoms map to different codes. Deep learning techniques suggested in literatures analyze these relationships and give precise prediction for new symptoms. For such deep analysis, it is necessary to represent the codes in proper manner to get useful information. Patients need to be represented using proper codes for giving inputs to design a supervised deep-learning EHR framework. Similar symptoms should be represented in the neighboring lower vector space. Such representations for deep learning analysis are reported in the literature [37]–[41]. Examples of medical codes for sample symptoms are given in [1] and represented in Table 6.2.

There are different representations, namely the representation based on concept, distributed, and latent encoding. The representation that identifies the similarities in the medical concepts and groups them together using an unsupervised technique is called the concept-based representation. In distributed embedding, natural language processing was used for creating a comprehensive and compact representation from the vast medical details and codes. When there are several codes with the same temporal details, it

Table 6.2　Example of medical codes

S. No.	Schema	Mode	Total codes	Code	Diseases
1	ICD-10	Diagnosis	68,000	J9600	Respiratory disorder
				I509	Heart problem
2	CPT	Procedure	9641	72146	Thoracic spine
				67810	Eyelid biopsy
3	LOINC	Lab test	80,868	4024-6	Salicylate, serum
				3414-0	buprenorphine screen
4	RxNorm	Drugs	116,075	161	Acetaminophen
				7052	Morphine

is dealt with by splitting codes into several small groups and then ordering within based on the occurrence of events. Skip-gram is generally used for such purposes [37], [42], [43]. In latent encoding, the correlations between different medical concepts are analyzed and linear models are trained with illustrations obtained through autoencoders [44].

Deep learning techniques [39], [40], [42], [43], [45] are used to represent patients in the form of vector based on natural language processing [46] or decrease the dimensionality using autoencoders [26]. Stacked three-layer AEs are designed in such a way that the final layer of weights provides the representation of patients. The patient's time data was also extracted to know whether the code identification was done at the earliest from EHR records.

6.4.3 Evaluation of representation

The representations are usually evaluated by classification techniques and the accuracy of evaluation depends on the presentation of clinical data in EHR in a perfect manner. Certain studies concentrate on evaluating the representation in a direct manner without any next-level categorization. So far, there are no common evaluation parameters for the output produced by deep learning networks for clinical data. Hence, the assessment could not be done for the prediction done by various deep learning networks. The common parameters used in literature are concept coherence [40], [43], medical concept similarity measure (MCSM) and medical relatedness measure (MRM) for evaluating EHR medical codes. Execution of deep learning EHR works are not done for common benchmark data sets by the researchers, and, thus, evaluation could not be done based on the results attained in their works.

6.4.4 Diagnosis of disease

The final aim of deep learning for EHR analysis is the diagnosis of disease based on the available records. The diagnosis could be done immediately when the patient's data shows the symptoms clearly or it can be delayed and further tests for clear prediction will be advised. Based on the timing of prediction, the diagnosis is categorized into static or temporal. Temporal prediction takes time for final diagnosis, while static prediction identifies the disease at the first encounter itself. Heart failure, hypertension, osteoporosis, etc., are categorized as static, whereas cardiovascular, renal diabetes, mental health, and some categories of heart failure are categorized as static. Unsupervised learning is the commonly used technique for such prediction.

The accuracy of such predictions could be enhanced by improving the representation of deep learning concepts. Temporal events are not considered for static prediction and researchers have proved that MLP is the best model among ANNs [37] for such prediction. The classifier designed for suicide risk identification [24] with the entire EHR data performs well compared to the classifier that used only diagnosis codes. Some deep learning architecture requires a certain time period to diagnose the disease. A CNN network with temporal medical details is used to predict heart problems and is found to work better with good results [47]. Skip-gram, GRU, and LSTM networks were also used in literature for the diagnosis and labeling of codes [39], [45], [48]. MLP and LSTM networks are used by Nickerson *et al.* [49] for the prediction of postoperative complications, whereas two more works suggest CNN and RNN for predicting re-admission after discharge from hospitals and for complications after kidney transplantation [38], [50].

6.4.5 Evaluation metric

The foremost issue of EHR data is the classification of different medical codes. The medical images need to be segmented or classified based on the deep learning strategies. The output of such classification should be evaluated by common metrics. Accuracy, precision, and recall are the commonly used metrics for rating the diagnosis output of deep learning networks. The classification results of existing literatures cannot be compared directly based on these existing results as there is no common benchmark data set.

6.4.6 Risk identification and survival prediction

In order to prevent the re-occurrence of a disease, the factors responsible for it should be identified properly. Such identification also assists the doctors to provide good treatment for the patients to recover early from the disease. Literatures also suggest the usage of deep networks to identify features from the risk factors stored in EHR data [51], [52].

6.5 EHR Data Set

The EHR data set is available with the hospitals and is difficult to obtain such data. Such data is available in various formats like image, text, numeric, etc. [8]. The results reported by deep learning research are based on proprietary data set and not available in public. Few of the data sets available in public are

Medical Information Mart for Intensive Care (MIMIC) and i2b2: Informatics for Integrating Biology and the Bedside [53], [54]. PCORnet database is made up of partner network – 13 clinical data research networks (CDRNs). PCORnet refers to the common data model (CDM) to regulate clinical records. The data world website https://data.world/datasets/ehr also contains EHR data sets which are used by several projects.

6.6 Research Gap

The conventional machine learning approaches used for EHR data analysis were now being replaced by deep learning techniques and are proved to produce good results. Even though the recent literature suggests the success of deep learning techniques in clinical data analysis, there is still a long way to go for accurate analysis and prediction which necessitated further studies and improvement of existing methodologies. The major problem is acquiring EHR data that is available with hospitals. For maintaining the privacy of their patients' data, most of the hospitals deny providing them for analysis by researchers [8]. As a result, the data available for practical analysis is very little despite the huge growth of EHR data in hospitals. Even though few data are available in public, the data did not cover all the medical complications, and, hence, it is tough to proceed with deep analysis with that little quantity of data. For designing a robust and generalized deep EHR model, large quantity of data is required. The second major problem is the highly unstructured and unorganized arrangement of EHR data with more noise, errors, and missing data [8]. These two major issues form bottlenecks in the design of deep EHR record analysis.

Another tough task is the assignment of medical codes based on symptoms that require the support of doctors or nurses. This deep EHR analysis is a collaborative work between researchers and medicos for the improvement in diagnosis and prediction that could help to serve the patients in a better way. Hence, the hospitals should be encouraged to share the data with the researchers, and, in turn, the researchers should not disclose the patients' details to the public. The clinical data in different formats like text, medical codes, image, scan reports, etc., pose serious problems and is very critical to carefully do the fusion of all such data and represent them in a common format for proper analysis. The data is more heterogeneous in nature as they are collected from different sources and there is no common representation of such data. These really pose a tough problem during the normalization process.

There is still a huge hesitation to apply the outcomes of deep learning research in real-time hospital environments due to the reasons cited above which necessitates further research in this area for practical deployment. A robust and efficient model that can analyze the data very fast with good accuracy is the need of the hour for a clinical decision support system. Hence, it is time for the medical staff and researchers to work together to provide such a good EHR data analysis model that works well in real time.

6.7 Suggestions for Improvement

It is hard to interpret the resulting models by using deep learning techniques. So, these techniques are called "black boxes". Barter between the rendition and frankness is created due to this quandary. Predictions and guessing might have affected correct diagnostics and treatments of the patients in the medical world.

That is why linear models like logistic regression are meant as best in the clinical domain. We are going to review trials to make this deep learning technique to be clearer. In [27], [48], and [55]–[57], the main limitation is lack of interpretability. In Table V, a summary of interpretability techniques is shown.

6.7.1 Tuning activation function

It is necessary to study the input types clearly as it will affect the activation of hidden and output units. Such a study will surely enhance the learning procedure. The significance of each and every feature should be identified to design a perfect deep learning model [38], [42], [47], [58].

6.7.2 Constraints

Numerous researchers have examined training limitations in order to improve the interpretability of deep models. As a result, training parameters like the number of epochs, the number of inputs, and the number of hidden layers should be precisely set. Similarly, weight adjustment should be focused appropriately [40], as the resulting weight sparsity reveals the types of inputs that activate the sparse neurons. All of these weight limitations should be incorporated into the goal function used to tune the architecture of the deep learning neural network. The [59] phenotypic discovery approach for time series data increases the interpretability of learned model weights by imposing irregularity and sparsity requirements.

6.7.3 Qualitative clustering

Natural clustering of vectored representations in the EHR concept can be accomplished with the use of a visualization approach such as t-stochastic neighbor embedding (t-SNE). This is actually a two-dimensional representation of pair-wise similarity between high-dimensional data points [55]. For deep learning to perform well, it is necessary to do a thorough examination of the phenotypic components. In the case of similar diagnoses and patient subgroups, a distributed representation [32] of medical events and patient vectors in two dimensions enables the greatest performance.

6.7.4 MIMIC learning

Deep model transparency should be addressed adequately in their interpretable MIMIC learning frameworks [59]–[60]. Interpretable linear models can be used to weigh the raw input features, significantly increasing the power of deep networks. Maintaining desired feature transparency while creating deep architecture is another critical consideration.

Conclusion

This chapter provides a review of current deep learning research, as it pertains to EHR analysis. Numerous publications on the burgeoning topic were published in the last two years. When we look back at the breakthroughs in image and natural language processing made possible by deep learning, we see a striking parallel to the domain of contemporary EHR-driven deep learning research. The majority of research in this survey demonstrates the most accurate technique to define the large volumes of raw patient data collected over the last decade. Basic image processing techniques are concerned with increasingly complicated and hierarchical representations of pictures. Similarly, natural language processing is critical, as most of the EHR data is composed of physician notes that must be properly understood. Future research should focus on approaches to describe patient health data and their unique medical codes so that the disparate sources of data do not create bottlenecks in EHR data analysis. This chapter discussed the clinical record analysis process, the deep learning networks that have been deployed so far in this, the research gap, and finished with some recommendations for future deep learning research enhancement.

References

[1] Benjamin Shickel , Patrick James Tighe , Azra Bihorac , and Parisa Rashidi, "Deep EHR: A Survey of Recent Advances in Deep Learning Techniques for Electronic Health Record (EHR) Analysis", IEEE Journal of Biomedical And Health Informatics, vol. 22, No. 5, pp 1589-1604, September 2018

[2] A. Subasi and M. I. Gursoy, EEG signal classification using PCA, ICA, LDA and support vector machines, Expert Syst. Appl., vol. 37, no. 12, pp. 8659–8666, 2010.

[3] M. Pechenizkiy, A. Tsymbal, and S. Puuronen, PCA based feature transformation for classification: Issues in medical diagnostics, in Proc. 17th IEEE Symp. Computer-Based Medical Systems, Bethesda, MD, USA, 2004, pp. 535–540.

[4] L. V. Lita, S. P. Yu, S. Niculescu, and J. B. Bi, Large scale diagnostic code classification for medical patient records, in Proc. 3rd Int. Joint Conf. Natural Language Processing, Hyderabad, India, 2008, pp. 877–882.

[5] B. Koopman, G. Zuccon, A. Nguyen, A. Bergheim, and N. Grayson, Automatic ICD-10 classification of cancers from free-text death certificates, Int. J. Med. Inform., vol. 84, no. 11, pp. 956–965, 2015.

[6] D. X. Wang, X. Liu, and M. D. Wang, A DT-SVM strategy for stock futures prediction with big data, in Proc. 16th Int. Conf. Computational Science and Engineering, Sydney, Australia, 2013, pp. 1005–1012.

[7] J. M. Sun and C. K. Reddy, Big data analytics for healthcare, in Proc. 19th ACM SIGKDD Int. Conf. Knowledge Discovery and Data Mining, Chicago, IL, USA, 2013, p. 1525.

[8] Ying Yu, Min Li, Liangliang Liu, Yaohang Li, and Jianxin Wang, "Clinical Big Data and Deep Learning: Applications, Challenges, and Future Outlooks", Big Data Mining And Analytics, Vol.2, No.4, pp288–305, December 2019.

[9] A. Perotte, R. Pivovarov, K. Natarajan, N. Weiskopf, F. Wood, and N. Elhadad, Diagnosis code assignment: Models and evaluation metrics, J. Am. Med. Inform. Assoc., vol. 21, no. 2, pp. 231–237, 2014.

[10] M. Li, Z. H. Fei, M. Zeng, F. X. Wu, Y. H. Li, Y. Pan, and J. X. Wang, Automated ICD-9 coding via a deep learning approach, IEEE/ACM Transactions on Computational Biology and Bioinformatics, doi: 10.1109/TCBB.2018.2817488.

[11] H. R. Shi, P. T. Xie, Z. T. Hu, M. Zhang, and E. P. Xing, Towards automated ICD coding using deep learning, arXiv preprint arXiv: 1711.04075, 2017.

[12] T. Baumel, J. Nassour-Kassis, R. Cohen, M. Elhadad, and N. Elhadad, Multi-label classification of patient notes a case study on ICD code assignment, arXiv preprint arXiv: 1709.09587, 2017.

[13] M. Zeng, M. Li, Z. H. Fei, Y. Yu, Y. Pan, and J. X. Wang, Automatic ICD-9 coding via deep transfer learning, Neurocomputing, vol. 324, pp. 43–50, 2019.

[14] S. Gehrmann, F. Dernoncourt, Y. R. Li, E. T. Carlson, J. T. Wu, J. Welt, J. Jr. Foote, E. T. Moseley, D. W. Grant, P. D. Tyler, et al., Comparing rule-based and deep learning models for patient phenotyping, arXiv preprint arXiv: 1703.08705, 2017.

[15] S. Gao, M. T. Young, J. X. Qiu, H. J. Yoon, J. B. Christian, P. A. Fearn, G. D. Tourassi, andRamanthan, Hierarchical attention networks for information extraction from cancer pathology reports, J. Am. Med. Inform. Assoc., vol. 25, no. 3, pp. 321–330, 2018.

[16] H. C. Shin, L. Lu, L. Kim, A. Seff, J. H. Yao, and R. M. Summers, Interleaved text/image Deep Mining on a large-scale radiology database for automated image interpretation, J. Mach. Learn. Res., vol. 17, no. 1, pp. 3729–3759, 2016.

[17] L. Dai, R. G. Fang, H. T. Li, X. H. Hou, B. Sheng, Q. Wu, and W. P. Jia, Clinical report guided retinal microaneurysm detection with multi-sieving deep learning, IEEE Trans. Med. Imaging, vol. 37, no. 5, pp. 1149–1161, 2018.

[18] F. Duarte, B. Martins, C. S. Pinto, and M. J. Silva, A deep learning method for ICD-10 coding of free-text death certificates, in Proc. 18th EPIA Conf. Artificial Intelligence, Porto, Portugal, 2017, pp. 137–149.

[19] H. Z. Fu, J. Cheng, Y. W. Xu, D. W. K. Wong, J. Liu, and X. C. Cao, Joint optic disc and cup segmentation based on multi-label deep network and polar transformation, IEEE Trans. Med. Imaging, vol. 37, no. 7, pp. 1597–1605, 2018.

[20] A. Fakhry, T. Zeng, and S. W. Ji, Residual deconvolutional networks for brain electron microscopy image segmentation, IEEE Trans. Med. Imaging, vol. 36, no. 2, pp. 447–456, 2017.

[21] D. M. Pelt and J. A. Sethian, A mixed-scale dense convolutional neural network for image analysis, Proc. Natl. Acad. Sci. USA, vol. 115, no. 2, pp. 254–259, 2018.

[22] M. Li, Z. H. Fei, M. Zeng, F. X. Wu, Y. H. Li, Y. Pan, and J. X. Wang, Automated ICD-9 coding via a deep learning approach, IEEE/ACM Transactions on Computational Biology and Bioinformatics, doi: 10.1109/TCBB.2018.2817488.

[23] S. Hochreiter and J. Schmidhuber, Long short-term memory, Neural Comput., vol. 9, no. 8, pp. 1735–1780, 1997.

[24] H. R. Shi, P. T. Xie, Z. T. Hu, M. Zhang, and E. P. Xing, Towards automated ICD coding using deep learning, arXiv preprint arXiv: 1711.04075, 2017.

[25] T. Baumel, J. Nassour-Kassis, R. Cohen, M. Elhadad, and N. Elhadad, Multi-label classification of patient notes a case study on ICD code assignment, arXiv preprint arXiv: 1709.09587, 2017

[26] I. Goodfellow, Y. Bengio, and A. Courville, *Deep Learning*. Cambridge, MA, USA: MIT Press, 2016

[27] E. Choi, M. T. Bahadori, A. Schuetz, W. F. Stewart, and J. M. Sun, Doctor AI: Predicting clinical events via recurrent neural networks, in Proc. 2016 Machine Learning for Healthcare Conf., Los Angeles, CA, USA, 2016, pp. 301–318.

[28] A. Khatami, A. Khosravi, T. Nguyen, C. P. Lim, and S. Nahavandi, Medical image analysis using wavelet transform and deep belief networks, Exp. Syst. Appl., vol. 86, pp. 190–198, 2017

[29] K. L. Hua, C. H. Hsu, S. C. Hidayati, W. H. Cheng, and Y. J. Chen, Computer-aided classification of lung nodules on computed tomography images via deep learning technique, Onco Targets Ther., vol. 8, pp. 2015–2022, 2015.

[30] S. Ebadollahi, J. Sun, D. Gotz, J. Hu, D. Sow, and C. Neti, "Predicting patient's trajectory of physiological data using temporal trends in similar patients:Asystem for near-term prognostics," in *Proc. AMIA Annu. Symp.*, 2010, pp. 192–196.

[31] D. Zhao and C. Weng, "Combining PubMed knowledge and EHR data to develop a weighted Bayesian network for pancreatic cancer prediction," *J. Biomed. Informat.*, vol. 44, no. 5, pp. 859–868, 2011.

[32] A.N. Jagannatha and H.Yu, "Structured prediction models for RNN based sequence labeling in clinical text," in *Proc. Empirical Methods Natural Lang. Process.*, 2016, pp. 856–865.

[33] A.N. Jagannatha and H. Yu, "Bidirectional recurrent neural networks for medical event detection in electronic health records," in *Proc. conf. Assoc. Comput. Linguistics North American Chapter Meeting*, 2016, pp. 473–

[34] J. A. Fries, "Brundlefly at SemEval-2016 Task 12: Recurrent neural networks vs. joint inference for clinical temporal information extraction," in *Proc. 10th Int. Workshop Semantic Eval.*, 2016, pp. 1274–1279.

[35] A.De Sa *et al.*, "Incremental knowledge base construction using Deep-Dive," *VLDB J.*, vol. 26, pp. 81–105, 2017.

[36] Y. Liu, T. Ge, K. S. Mathews, H. Ji, D. L. Mcguinness, and C. Science, "Exploiting task-oriented resources to learn word embeddings for clinical abbreviation expansion," in *Proc. 2015 Workshop Biomed. Natural Lang.Process.*, 2015, pp. 92–97.

[37] E. Choi, A. Schuetz, W. F. Stewart, and J. Sun, "Medical concept representation learning from electronic health records and its application on heart failure prediction," arXiv:1602.03686, Feb. 2016, p. 45.

[38] P. Nguyen, T. Tran,N.Wickramasinghe, and S.Venkatesh, "Deepr: A convolutional net for medical records," in *IEEE J. Biomed. Health Informat.*, vol. 21, no. 1, pp. 22–30, Jan. 2017.

[39] T. Pham, T. Tran, D. Phung, and S.Venkatesh, "Deep Care:A deep dynamic memory model for predictive medicine," in *Pacific-Asia Conf. Knowl. Discovery Data Mining*, Springer International Publishing, Apr. 2016, pp. 30–41.

[40] T. Tran, T. D. Nguyen, D. Phung, and S. Venkatesh, "Learning vector representation of medical objects via EMR-driven nonnegative

restricted Boltzmann machines (eNRBM)," *J. Biomed. Informat.*, vol. 54, pp. 96–105, 2015.

[41] S. Mehrabi *et al.*, "Temporal pattern and association discovery of diagnosis codes using deep learning," in *Proc. 2015 Int. Conf. Healthcare Informat.*, 2015, pp. 408–416.

[42] E. Choi,M.T.Bahadori, E. Searles,C.Coffey, and J. Sun, "Multi-layer representation learning for medical concepts," in *Proc. 22nd ACM SIGKDD Int. Conf. Knowl. Discovery Data Mining*, Aug. 2016, pp. 1495–1504

[43] Y. Choi, C. Y.-I. Chiu, and D. Sontag, "Learning low-dimensional representations of medical concepts methods background," in *Proc. AMIA Summit Clin. Res. Informat.*, 2016, pp. 41–50.

[44] X. Lv, Y. Guan, J. Yang, and J.Wu, "Clinical relation extraction with deep learning," *Int. J. Hybrid Inf. Technol.*, vol. 9, no. 7, pp. 237–248, 2016.

[45] E. Choi, A. Schuetz, W. F. Stewart, and J. Sun, "Using recurrent neural network models for early detection of heart failure onset," *J. Amer. Med. Informat. Assoc.*, vol. 292, no. 3, pp. 344–350, 2016.

[46] T. Mikolov, G. Corrado, K. Chen, and J. Dean, "Efficient estimation of word representations in vector space," in *Proc. Int. Conf. Learn. Representations*, 2013, pp. 1–12.

[47] Y. Cheng, F. Wang, P. Zhang, H. Xu, and J. Hu, "Risk prediction with electronic health records: A deep learning approach," in *Proc. SIAM Int. Conf. Data Mining*, 2015, pp. 432–440.

[48] Z. C. Lipton, D. C. Kale, C. Elkan, and R.Wetzell, "Learning to diagnose with LSTM recurrent neural networks," arXiv:1511.03677, 2016.

[49] P. Nickerson, P. Tighe, B. Shickel, and P. Rashidi, "Deep neural network architectures for forecasting analgesic response," in *Proc. 2016 IEEE 38th Annu. Int. Conf. Eng. Med. Biol. Soc.*, 2016, pp. 2966–2969.

[50] A. Esteban, O. Staeck, Y. Yang, and V. Tresp, "Predicting clinical events by combining static and dynamic information using recurrent neural networks," in *Healthcare Info. (ICHI), 2016 IEEE Int. Conf.*, Oct. 2016, pp. 93–101.

[51] R. Poplin, A. V. Varadarajan, K. Blumer, Y. Liu, M. V. McConnell, G. S. Corrado, L. Peng, and D. R. Webster, Predicting cardiovascular risk

factors from retinal fundus photographs using deep learning, arXiv preprint arXiv: 1708.09843, 2017.

[52] H. Li, X. Y. Li, M. Ramanathan, and A. D. Zhang, Prediction and informative risk factor selection of bone diseases, IEEE/ACM Transactions on Computational Biology and Bioinformatics, vol. 12, no. 1, pp. 79–91, 2015.

[53] F. Dernoncourt, J. Y. Lee, O. Uzuner, and P. Szolovits, "De-identification of patient notes with recurrent neural networks," *J. American Medical Informat. Assoc.*, vol. 24, no. 3, pp. 596–606, 2017.

[54] A. E. Shweta, S. Saha, and P. Bhattacharyya, "Deep learning architecture for patient data de-identification in clinical records," in *Proc. Clin. Natural Lang. Process. Workshop*, 2016, pp. 32–41.

[55] R. Miotto, L. Li, B. A. Kidd, and J. T. Dudley, "Deep patient: An unsuper-vised representation to predict the future of patients from the electronic health records," *Sci. Rep.*, vol. 6, no. Apr., 2016, Art. no. 26094.

[56] L. Nie, M. Wang, L. Zhang, S. Yan, B. Zhang, and T.-S. Chua, "Dis-ease inference from health-related questions via sparsely connected deeplearning," *IEEE Trans. Knowl. Data Eng.*, vol. 27, no. 8, pp. 2107–2119, Aug. 2015.

[57] H. I. Suk and D. Shen, "Deeplearning-based feature representation for AD/MCI classification," in *Proc. 16th Int. Conf. Med. Im-age Comput. Comput. Assisted Intervention*, 2013, vol. 8150, pp. 583–590.

[58] Z. Che, D. Kale, W. Li, M. Taha Bahadori, and Y. Liu, "Deep computational phenol typing," in Proc. 21st ACM SIGKDD Int. Conf. Knowl. DiscoveryDataMining, 2015, pp. 507–516.

[59] G. J. Kuperman *et al.*, "Medication-related clinical decision support in computerized provider order entry systems: A review," *J. Amer. Med. Informat. Assoc.*, vol. 14, no. 1, pp. 29–40, 2007.

[60] K. P. Murphy, *Machine Learning: A Probabilistic Perspective*. Cambridge, MA, USA: MIT Press, 2012.

7

Telemedicine-based Development of M-Health Informatics using AI

Jagjit Singh Dhatterwal[1], Kuldeep Singh Kaswan[2], and Naresh Kumar[3]

[1]Department of Computer Applications, PDM University, India
[2]School of Computing Science and Engineering, Galgotias University, India
[3]School of Computing Science and Engineering, Galgotias University, India

Abstract

One of the main problems in m-Health is how to effectively utilize the mobile telecommunications capabilities that are now nearly universally available. The purpose of this article is to demonstrate a multi-channel m-Health system with a Bluetooth connection that is based on the general packet radio service (GPRS). The goal here is to create a system that uses a smart telephone on a commercially GPRS network to send a patient's biological signals immediately to a hospital. To observe and handle health information, the system is coupled with client computer applications. The telemedicine session is handled by an application server, which also controls the client request message from a distant patient. All diagnostic information transferred during a telemedicine session, along with patient identification and telemedicine conversation detailed information, is saved in a database for later analysis. Clinicians can obtain these data as needed by using browser applications to query the repository.

7.1 Introduction

The conventional method of delivering telemedicine technology was to send biological signals from a patient to a clinician using "landline" communication technologies such as the public switched telephone network (PSTN) and the integrated services digital network (ISDN). In radiology, cardiology, pathology, and surgeries, transmissions encoding parameters such as the electrocardiogram (ECG), hypertension, temperature, and oxygen levels at low bandwidths and X-ray pictures and ultrasound imaging at high available bandwidth have been sent [1]. Telemedicine has emerged as a quick and rapidly developing topic of research due to the concept of giving clinical records and hospital service providers to a faraway patient. Telemedicine has significant potential to improve the quality and availability of hospital facilities in both towns and cities, as well as in industrialized and developing economies, because it means allowing healthcare workers to be supervised in a genuinely remote manner, possibly while on the move and from nearly anywhere around the entire globe. The limits of traditional telemedicine are self-evident because operations are confined to conversations between fixed sites, frequently using windows software. As a result, combining mobile cellular connections with telemedicine apps is seen as the next critical step toward providing more mobile and cost-effective healthcare services. Interaction between technicians and physicians is beginning to show that this combination is both practicable and beneficial for healthcare delivery. Telemedicine has benefited greatly from advances in wireless communications technologies, which range from permanent technology platforms to mobile communication capabilities. Furthermore, simultaneous advancements in ubiquitous, highly advanced, and wearable systems have given rise to the m-Health idea, which is the 21st-century overarching concept telemedicine. m-Health may be described broadly as "mobile computer, medical sensor, and telecommunication technology for universal healthcare". Although face-to-face discussions between a physician and a patient should never be eliminated, there are some medical problems that can be treated more effectively utilizing m-Health technologies. Remote regular check-ups, disaster response scenarios, and physiologic measures in sports science are all possible software apps. Medical services may now be supplied to any locations within the communication range of wireless networks like the global system for mobile communications (GSM). The ability to monitor individuals efficiently and consistently while in a hospital provides a significant potential to reduce the strain on national healthcare resources. An individual in a rural region may receive a routine check through mobile phone instead of needing to go to a hospital on a constant schedule. Frequent checks and surveillance can be performed while the individual is at home, on the road, at work, or at play, freeing up hospital personnel for more difficult patients. This could save both doctors and patient's effort and cash, particularly by eliminating a need for the individuals to travel, frequently for short appointments after long delays [2].

7.1 Objectives of Chapter

The goal of this study is to create an incorporated Bluetooth m-Health system for transmitting multi-channel sensor data via one or more wireless communication systems. The following are the study's objectives:

- To develop and construct a modularity m-Health processing unit to sample, process, and communicate biomedical signals
- To create a Bluetooth communications channel between the m-Health processing and a cell telephone
- To send sampled biomedical signals from a patient to a hospital via a cell connection and study the network efficiency in diverse mobile settings

7.2 Literature Review

This section focuses on literature that uses key management approaches to protect medical data in healthcare sensor networks.

Sriram Cherukuri et al. (2003) suggested a new way to dealing with medical sensor networks security concerns. This system uses biometric obtained from the female organism to safeguard the sensitive data needed to protect the transfer of data. In order to secure the key for measuring errors and a randomization issue, one solution uses error corrections and various biometrics. This method removes the necessary calculations and dramatically lowers communication in comparison to conventional asymmetric key setup procedures. In the future, important biometric data are collected and their temporal differences examined and a combination of biometrics is developed that will result in adequate randomization.

Shu-Di Bao et al. (2005) suggested an EA system to address the issue of body identification in m-body health's area sensor network (BASN). Twelve healthy people with 2 photoplethysmograms were studied on a suggested entity authentication system (PPG) captured at several areas of the body synchronously. As a biometric feature, the beat-to-beat cardiac range is utilized to create individual identification. This is an advanced and safe system model during the establishment of wireless links and shown to be the best solution comparing with general encryption computing and minimal memory requirements. There is a potential future to use the chaotic character of the research participant for encrypted communication. In order to implement the suggested method in practice, further research work should be undertaken integrating the regard to sustainable development.

Patrick Traynor, Guohrong et al. (2006) introduced deterministic pairwise key-predistribution scheme (DPKPS) for medical sensor networks. In four steps, DPKPS is performed, namely initial installation, use, discovery, and key setup. In comparison with fully paired key schedule systems, this approach allows powerful BSN security solutions for extremely low power and memory cost. DPKPS increases the scalability of KPS, enhances the scalability of the x n-KPS, and enhance its resilience. In addition, the study will be expanded to fulfill the key BSNs management infrastructure in the medical field, involving the creation of effective revoking, updating, and controlling important materials in BSNs.

Carmen C. Y. Poon and Yuan-Ting Zhang (2006) suggested a unique biometric approach to securing a telemedicine and M-based wireless body networking. This article looks at a biometric method that employs an innate property of the human body to ensure an identification or redistribution of chips to secure the transmission of the intersensory network. In 99 participants with an 838 segment simultaneously recording ECG- or PPG-symbols, this approach employed inter-pulse interval (IPI) as the biometric characteristic and test. In comparison to the authentication method (Shu-Di Bao et al. (2005)), this biometric approach gives a higher degree of security with less memory and less calculation. This work has highlighted several critical questions for further study, including remedial methods for various channel asynchronies owing to illnesses, physiological events, motion artifacts, dictionary mistakes, etc.

Shu-Di Bao et al. (2008) proposed a new light weight biometric technique that utilizes intrinsic characteristically character traits of a human body to produce an entity identification (EI) of each of the BSN to recognize a node and safeguard the transmission of important inputs to address security problems and overcome the severe resource constraint challenges in BSN. This approach mainly employs EIs to secure intra-BSN communications. It shows the interpole interfaces (IPIs) of the ECG and photothysmogram of 99 patients computed from the heart rhythm.

In comparison with the biometrics Upkar Varshney (2008) introduced a resource-friendly key management system in body sensor networks for the generation and distribution of cryptographic keys and suggested an alternative for symmetrical encryption algorithms for certain types of data to the new methodology for data clamoring based on extrapolation and accidental selection. The system may be modified to meet the necessary resource limitations depending on the specific application. Adjustable parameters for various levels of resilience and complexity exist for the techniques provided. The methods proposed compared with the conventional single point fuzzy key approach this system makes more optimal use of transport and resource usage. But in many real circumstances, such as canal fading and aberrations, the resilience of the system has to be examined.

The new key agreement technique, EKG-based key agreement (EKA), introduced by VeikkoIkonen and Eija Kaasin et al. (2008), which employed electrocardiograms (EKGs) for the creation of cryptographic keys. The fast Fourier transformer extracts the feature. Using EKA, a "fitting-n-play" way may be used to ensure securing communication between the sensors. No system setup was necessary, and the body received keys for communication as and when necessary. In comparison to the technique given by L. Biel et al. (2001), the authors offered efficient process for creating EKG template. Finally, the authors set forth their planning of future work to increase the security of the EKA scheme by broadening the block size, using actual hardware techniques, evaluating the key developed in greater detail, and studying the computer and waste implementation of the method. Vijay Srinivasan, John Stankovic, and Kamin (2008) presented an impoverished EKG basic agreement approach that uses discrete wavelet transformation (DWT) to provide a shared key inside a body area network. The benefit of DWT over FFT is its linear computing costs. This procedure maintains the security of the blocks when they are locked and unlocked between communication sensors with iris or fingerprinting. The system is secured using watermarking for trying to open. This system improves on the key agreement technique for safe and efficient Wi-Fi (SEKEBAN) based on the EKG.

Young-Dong Lee and Wan-Young Chung (2009) use WBAN-security physiological signals (ECG). SEKEBAN maintains and protects privacy by generating and distributing symmetric cryptographic keys to WBAN component sensors. Compared with commodity key generation (FrancisMinhthang Bui and Dimitrios Hatzinakos (2008)), SEKEBAN is efficient and energy efficient. To guarantee end-to-end communication, it seeks to create safe and effective symmetric public key between both the sensor nodes of the network. It also seeks to safeguard communication relationships with biometric data among the sensing nodes individually.

Eryun Liu et al. (2010) have presented the useful and effective approach to secure body sensor network histocompatibility systems for physiological safety values (PVS). The PVS system distributes, by concealing physiological data, the key required to secure a message together with the information itself. This approach minimizes the need for intentional key exchange to satisfy all its safe communication criteria by reducing the number of keys necessary at each node. This system was energy-saving, space-effective, and more dynamic than LEAP protocol comparing to safety in a body sensor network

(Zhu et al. 2006). In the future, the authors will explore approaches for the elimination of topographic specificities that are capable of entirely removing the main distribution.

Krishna K. Venkatasubramanian et al. are proposing highly flexible authentication and key establishment procedures based on the ECG signals and fuzzy commitment, namely the ESKE (2010). The singularity of ECG signals ensures that ESKE offers lengthy, random, low latency, identifiable, and temporally variant keys. ESKE employs the same fuzzy PSKA engagement method, which can handle the high noise level and fluctuation of ECG data. But for feature generation, ESKE employs discrete wavelet transforms. ESKE does not need a prior dissemination of the key content. The test findings demonstrate, therefore, that ESKE improves security when compared to PSKA with regard to data secrecy, authentication, consistency, and harmful assaults.

In the wireless public health system, Lifang Wu et al. (2011) introduced authorship of biomes with physiological characteristics. This suggested technique for data authentication using electrocardiogram (ECG) signal patterns to reduce key overhead exchanges. The interpoles interval (IPI) in this method. The broadcaster side of the signal pattern is summarized as a biometric key with the Gaussian model of mixing (GMM). A lightweight signature checking system is employed on the receiving side, which leverages IPI signals locally collected on the receiver. Stochastic pattern recognition to ECG safety, which is essentially different from conventional techniques, is the main contribution of this suggested methodology. This method also benefits from a low HTER verification threshold for sample misaligned.

Min Chen et al. (2011) developed wireless body area network's trust key management scheme that uses biomedical parameters such as EKG to deal with WBAN security problems. This technique manages symmetric ECG cryptographic keys to be generated and distributed across sensor nodes safely and effectively via the WBAN to WBAN sensors. In comparison with SSL, this system efficiently saves energy.

The system termed ECG improved Jules Sudan (IJS), which uses the use of biomedical parameters such as the ECC, has been introduced by Zhaoyang Zhang et al. (2012). Without overhead redistribution, this key establishment technique may ensure data transmission over BANs in plug-n-play. To create keys with minimal computer requirements, ECG-IJS just needed a brief timeframe.

Daojing He et al. (2013) presented the secure wireless body area network (LBP) based hybrid kind of key management approach. The use of LBP to produce electrocardiogram signal makes this approach effective.

The use of SHA-256 on EKG blocks and important agreements to create blocks for change, disclosure, and secrecy of data provide security. Safety is guaranteed.

Daojing He et al. (2014) recognized the security liability for medical data discovery and suggested a WBAN protocol on secure information discovery and dissemination that is anonymous and denial-of-service to ensure the dissemination of the data is not interfered with. The authors modified each packet's secret key to prevent duplicating the keys at the halfway nodes. Since each packet requires this protocol to calculate the secret key, overhead computation will affect the spread of medical information from source to destination. This approach is created to defeat symmetrical cryptographic strategies and applications based on public key cryptography.

The biometric-based main agreements method, known as the PFKA, has been presented by Najeh Jammali and Lamia Chaari Fourati (2015) to provide secure communication between sensors. This approach enables sensors of the same WBAN, produced from overlapped medical image characteristics, to agree on a symmetrical decryption system. The examination of the performance of this system showed calm calculation costs, low storage, and a soft OVFKA communication.

A secure ECG-based biometric authentication scheme in the body area sensor networks was developed by Hussein Moosavi and Francis Minhthang Bui (2016). The authors showed how the ECG sensor chip and its management platform may be designed and implemented. The resultant sensor system is the first BAN sensor system described for real-time ECG, IPI, and authentication information of internodes. The descriptive statistics of actual sensor readings is important to the proposed authentication procedure, which enables the developer to modify system settings in line with the real-world situations BAN installations' features. A design process is performed, and each design phase discusses the strengths and limits of each step. Compared to ESKE, PSKA, and OPFKA, this protocol is better. Future study consists on extended protocol validation for more people in different scenarios and potential aberrant ECG characteristics.

Strong biometric faster and more effective authentication (PPG) technique for wireless body area networks and e-health application has been given to Tilendra Choudhary and M. Sabarimalai Manikandan (2016). The four phases followed in this approach are pre-processing, systolic peak identification, combination average pulsative waveform mining, and commonalities that meet a normalized correlation metric. The method is tested and confirmed using actual and pretending participants' PPG signals. Compared to machine-based PPG biometrics, this approach is straightforward.

Results indicated that the approach suggested is very latent in networks of wireless body areas. This technique does not address the potential for attacks and their attack effectiveness.

7.3 Wireless Technologies in m-Health

The idea of objective is to advance has been widely researched during the previous few decades. Most biological signals, such as ECG, temperature, and blood pressure, have consistently been communicated from humans and animals. Einthoven's monitoring and distribution of an ECG over 1.5 km through the public telephone line in 1903 was a significant milestone in the early stages of development of biotelemetry. Since its inception by S. R. Winters in 1921, radio communication has become the primary technology for biotelemetry. Early improvements were confined to a relatively brief, single-channel transceiver module. Bio-telemetry technology was also considered for use in mobile clinical emergency systems, although it was seldom utilized with portable devices. Significant advancements in biotelemetry spurred the use of wireless connectivity in telemedicine. Since then, the capacity to monitor patients from rural health centers and distant places has developed quickly as part of m-Health applications [3].

Wireless transmission technology advances have enabled innovations in m-Health systems more intriguing in recent years, adding to the critical "medical connection" sector of new m-Health applications. The original bio-telemetry idea evolved into m-diagnostic, surveillance, and on-the-go interactions among sick people and healthcare professionals with m-Health. In general, m-Health systems may be divided into three categories based on the techniques used: (i) satellite linkages, (ii) "short-range" networks and links, and (iii) mobile wireless connections (GSM, GPRS, and 3G). Many recent research works have focused on these technologies and how their incorporation into m-Health apps contributes to improved quality, higher productivity, and a more adaptable system.

The use of satellite technology enables telemedicine applications to be carried out in remote and inaccessible places. A telemedicine system with a satellites connection may be made publicly available in nearly any region of the world, even the most inaccessible and dangerous settings, such as Mount Everest. This aspirational and intelligent telemedicine project used a satellite phone to track the position, respiratory rate, ambient heat, temperature changes, and exercise habits of mountaineer son Mount Everest all through explorations and tried to claim to be the first to completing tasks real time monitoring of blood on a hospital outpatient person in truly remote or potentially dangerous conditions. During a high-altitude climbing trip to Mount Logan, signal boosters were also utilized to send echo cardiogram pictures, ECG tracing, and blood pressure data. Another satellite-related initiative has been the creation of a web-based picture archiving and communications system (PACS). The goal of this experimental system was to enable clinicians to search for and obtain condensed medical pictures from cloud computers. A web-based PACS would allow tiny hospitals in remote regions to quickly contact a central PACS and consistently obtain important picture data [4].

7.3.1 Wireless medical sensor technologies

Non-invasive biomedical sensors on the market today range from piezoelectrical components to infrared sensor to opto-electronic sensing devices. As short-range wireless and biotechnological technology gets better, the use of wireless medical sensors combined with networking capabilities has increasingly been adopted into digital health technologies. Using smart technology with wireless connection, a wireless body area network (WBAN) may now be created to gather biomedical parameters from diverse body regions. The deployed short-range connectivity technologies vary from RF transceivers to wireless headphones and the recently launched ZigBee standard. The wireless connection was used in this study because of its rising importance in wireless sensor applications. The notion of a wireless sensor network has been studied not just in healthcare applications but also in smart buildings, security, advanced robotics, and the automobile sector. As microelectronics and nanoelectronics advance, tiny biological sensors may be combined with ubiquitous carriers. Removable carriers might be a finger ring, a smart shirt, a bio-cloth T-shirt, or a knitting fabric made of fiber and yarn. All incorporated sensors have a communications gateway for transportable wide area connectivity. Telecommunication capabilities are also mentioned as a key concern in the existing and future difficulties of wearing technology health systems. Short- and long-range wireless and mobile telecommunications are used to connect sensors as well as a wearable technology system and a healthcare practitioner. As a result, the portable system established in this study should provide to the foundation of present and future progress in wearable electronics. As the population is aging and related illnesses grow more prevalent, society becomes more increasingly health-conscious, and individuals become "health consumers" seeking better health administration. People's perceptions of healthcare are changing away from traditional, care facility care, and toward customer care. It is anticipated that m-Health systems combined with remote monitoring would have a significant impact on changing current healthcare administration methods and ultimately to the supply of tailored upcoming medical applications [6].

7.4 Signals for Biomonitoring

The number and quality of medical studies that can be sent via m-Health applications varies. The signals used in the study are physiological markers that may be used for surveillance and routine inspection. The electrocardiogram (ECG), temperatures, pulse rate, and oxygen levels (SP02) are typical indications examined, with the ECG posing the most difficulty due to its complicated structure.

- **Electrocardiogram (ECG) Signals:** The ECG is one of the electrodermal signals generated by electric currents related with muscles and nerve cell movement in the human body. Electrical changes can be monitored by putting 3-lead or 12-lead ECG electrodes on the human skin of something like the chest or extremities, according to whether the sensors are used for surveillance or diagnosing cardiologist. The form and amplitude of the observed signal are altered based on electrochemical configurations. Each lead provides distinct information dependent on its exploring different to the heart's axis. In cardiology, ECG signals are routinely analyzed using stationary recordings for patients with advanced disease and portable monitor for mobility patients. In smart institutions and households, computerized information from mobile patients are relayed through a short-range radio connection to the nearest metal fence receiver and then over a LAN to a central management workstation [7].

- **Blood Pressure:** The ECG is one of the vibrational signals generated by electrical changes related with muscle or nerve cell movement in the internal organs. Electrical abnormalities can be monitored by putting 3-lead or 12-lead ECG electrode on the surface of the skin of the chests or limbs, dependent on whether the sensors are used for screening or diagnosing cardiology. The form and intensity of the observed signal are altered based on electrochemical configurations. Each lead provides distinct information dependent on its exploring different to the heart's axis. Because the heart functions as a four-chambered pumping for the systemic circulation, the cardiac cycle's pounding function represents diastole and diastolic stages. The cardiac cycle phase is when the heart is at rest or filling, and the systole phase is when the heart is beating or pounding [3.6]. The aorta pumps blood during each contraction of the heart, converting energy into pressures. The blood pressure determined during these contractions is referred to as systolic because it represents the highest pressure. During the requirements and evaluate of the heart, the aortic valves shut down to prevent blood from re-entering the heart. The remaining pressure in the vessels is then referred to as systolic pressure, which correlates to the minimum amount. Non-invasive blood pressure (NIBP) is among the most popular methods of measuring an oxygen saturation pressure utilizing an electrical monitor equipped with a pressure gauge.

- **Body Temperature:** The temperature is utilized to offer information about a patient's overall health. Normal internal body temperature ranges from 36.6 to 37.3 °C. A fever, defined as a normal body temperature above 37.3 °C, is the most frequent disease identified by normal body temperature. The elevated body temperature that occurs during a fever is caused by the body's response to an illness or sickness. Effects on the human body that may occur because of large variations in normal body temperatures.

7.4.1 Wireless communication for biomonitoring

L. M. Ericsson of Sweden created Bluetooth in 1994, and it was called after Harald Ridiculously obvious "Bluetooth" II, King of The Dames from 940 to 981 A.D. Bluetooth technology is a short, low-power radio technology that operates in the unlicensed 2.4-GHz commercial, academic, and medical (ISM) frequency spectrum. Its maximal data transfer rate and range possibilities, as well as data for other relatively brief wireless communication technology. Some of the most important advantages of short-range electromagnetic technology in general are also discussed. The ability to integrate Bluetooth transmitter module into a variety of products has encouraged the usage of this technologies in investigation. Bluetooth allows portable equipment to communicate wirelessly to voice and data applications via point-to-point or moment in time communications [8].

7.5 Telemedicine Application Server

In general, the telemedicine applications server's primary role is to manage telemedicine sessions in an m-Health system. The server manages the client request message from a patient's user experience application and replies appropriately to acknowledge the request and create a secured point-to-point communication with the patients. The telemedicine software system is also in charge of communicating with the MySQL telemedicine host computer and doing the necessary queries during a telemedicine session. These include recording metadata stored, such as the timestamp, as well as generating a database entry to keep the multi-channel healthcare data provided to the hospital. In addition, the software system provides interfaces for healthcare practitioners to monitor and administrate the m-Health system. Adding a new patient to the system, changing, or removing current patient information and tracking the quantity of events done by a patient are all administration activities. The telemedicine enterprise system, as mentioned in the preceding part on the development platform, is built with Visual Basic coding and the ADO database management system, allowing interaction with the MySQL Telemedicine Databases via an ODBC information API. The Winsock web programming interface is used to construct the relationship with the customer, which allows communications via the TCP transport layer protocol. It outlines the Telemedicine Implementation Server's functionality.

7.5.1 Server protocols

The protocols used by the Telemedicine Client Application are divided into three different routines: packet extractor, telehealth connection formation, and data uploading into the videoconferencing database. The server initially requests the user login information for the MySQL telemedicine database server as well as the device on the network. When the servers comprehend the first data message after establishing a connection between the clients, the package extracting program commences. The hospital ID and an IMSI identification would be included in the initial package from the customer program, which is being used to assess the person to the rest of the network. The next step in the package extracting procedure is to decode the receiving multiplexed multimedia health information and store the data from each broadcaster in the following composition. The data stored in each buffer is later utilized in the data upgrade process. When a patient finishes a telemedicine session, the server checks for the end of transmitted character and prepared to listen for another communication session.

7.5.2 Server graphical user interface

The telemedicine software system has four distinct interfaces that correlate to the server's four primary tasks, namely the login procedures, active connectivity management, session list, and patients' records. The initial interface is the login procedure, which occurs when a user (for example, an authorized medical staff member) starts the computer and is prompted for the username and password, password, host pseudonym for the telemedicine database (in this case, the Telemedicine Centre), and the host's configured port number. The user then selects the log-in option, after which the next active connections processes more effective appears [9].

7.6 Interface Program

The patient user experience programming allows the patient to undertake data acquisition from the telemedicine processing module. In general, the patient curriculum will interpret the multi-channel health information from the central processing unit via the wireless link and transfers the information to the telemedicine database server via either a World Wide Web backbone (a LAN, dial-up Internet, or Internet connectivity access from a network operator) or the GPRS network. Upon a satisfactory computer network, the outpatient application also communicates the patient ID issued by the institution. This allows the hospital's telephony software system to begin the process of storing information from the teleconference session in the repository.

7.6.1 Patient interface

In theory, the portable phone's client application reads multidimensional health information from the telemedicine processing module via its Bluetooth link. The data is subsequently transmitted by the cellular telephone to the hospital's telemedicine application server over the GPRS network. When contrasted to the patient application operating on a laptop, this setup provides more versatility and simplicity of use. However, owing to processing and hardware limitations and a considerably lower screen resolution, the mobile telephone patient program offers less software application than the patient programmed on a computer. The two first setup processes necessary on a notebook connection, wirelessly and wide area connectivity, are conducted after the patient begins to run the mobile telephone client software, rather than it was before. The software begins by connecting over wireless headphones to the telemedicine processing module. The next step is to leverage the mobile phone's Internet connectivity by designating an entry point that will also be utilized by the communication line to finish the transaction. The mobile telephone outpatient software is configured to interact with the patient's telemedicine database server by specifying the server's IP address as the destination node. After establishing the client-server point-to-point communication, the smartphone delivers the health information to the institution. While being transferred to the institution, the multi-channel material from the telehealth processing element is also shown on the monitor.

7.6.2 Doctor browser interface

The medical browsing interfaces application will enable physicians to see and maintain the data from telemedicine conversations on an as-needed basis. Clinicians can use the browser software to get health information, session knowledge, and real clinical data from the videoconferencing databases. Specialists may also be granted admin rights to the healthcare database, allowing them to add, remove, or amend content. The browser application may be used with a desktop or portable hard drives having network connections, such as a cell phone or a PDA. While on the go, doctors can access the browser software via a fixed-IP network (a LAN, dial-up broadband, or Internet connections from a network operator) or the GPRS network. The doctor browser interfaces program's system design. The browsing software created for the m-Health system includes two user interfaces: one for a laptop-based programmer and one for a mobile-phone-based program.

7.7 Experiment Work

This dataset is taken from Google trend. Most of the country provides telemedicine services. Singapore country completely depends on the telemedicine services save for seriously infections. Below 20 many countries less believe in telemedicine services in Table 7.1.

Table 7.1 Telemedicine services

Country name services is giving	Percentage of telemedicine services
Singapore	100
Philippines	90
United States	69
Indonesia	51
Kenya	50
Nepal	44

Table 7.1 Continued

Country name services is giving	Percentage of telemedicine services
Ghana	40
Pakistan	39
Nigeria	36
India	35
United Arab Emirates	35
Lebanon	34
Hong Kong	29
Canada	29
Thailand	27
Israel	24
Vietnam	23
Switzerland	20
South Africa	20
Sri Lanka	20
Malaysia	19
Saudi Arabia	18
South Korea	16
United Kingdom	12
Egypt	12
Ireland	10
Morocco	9
Australia	8
Taiwan	8
Romania	6
Sweden	5
Iran	4
Netherlands	4
Russia	4
Italy	4
Ukraine	4

Table 7.1   Continued

Country name services is giving	Percentage of telemedicine services
Colombia	4
Argentina	4
Peru	4
Chile	3
Brazil	3
Turkey	3
France	3
Poland	3
Spain	3
Germany	2
Mexico	2
Japan	2

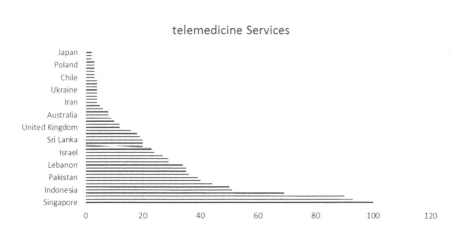

Figure 7.1   Telemedicine services in various countries.

Figure 7.1 shows the business of telemedicine in different countries. Japan is less interested in these services and Singapore prescribed medicine to people through tele-services.

Conclusion

To analyze and handle medical data, the m-Health system is linked with client-server applications. The telemedicine software system oversees managing the videoconferencing session and supervising the patient's client request message. Medical data transferred during a telemedicine session, as well as participant identification and telemedicine conversation information, are maintained in a database. Patient user experience applications are created for the client side to accept digitizing health information from the telemedicine central processing unit and transmit it to the healthcare server through the network connection. The patients' programmers are accessible for use on both a desktop and a mobile phone. The professional browsing applications enable doctors to access and monitor health information as well as patient information from the databases. The browser software on a laptop provides physicians with many more control and alternatives than the restricted surfing capabilities of the browser program on wearable electronics. The browsing application may be utilized with either a fixed network at the hospital or a phone network when on the road. A comparison of cordless security and protection illustrates the screening procedures offered. The protection measures included into the m-Health system serve to safeguard medical data consistency and patients' medical confidentially.

References

[1] Pronab Ganguly and Pradeep Ray, Software Interoperability of Telemedicine Systems, IEEE CNF 4-7 July 2000, available on [http://ieeexplore.ieee.org/iel5/6934/18631/00857717.pdf?tp=&arnumber=857717&isnumber=18631]

[2] Cameron T. Howie, et al; Software Interoperability; Center for Integrated Facility Engineering, Stanford University, November, 1996; available on [http:// www. dacs.dtic.mil/techs/interodtitle.shtml]

[3] Supavadee Aramvith and Ming-Ting Sun, MPEG-1 AND MPEG-2 Video Standards, 1999 Academic Press, available on [http://www.ee.eng.chula.ac.th/~supava/doc/mpeg12.pdf]

[4] J. R Epps and W.H. Holmen, A New Very Low Bit Rate Wideband Speech Coder with a Sinusoidal High band Model, IEEE, 2001 available on [http://ieeexplore.ieee.org/iel5/7344/19923/00921079.pdf]

[5] P. Venkat Rangan and Daniel C. Swinehart, Software Architecture for Integration of Video Services in the Etherphone System, IEEE Journal In Communications. Vol 9. No 9. December 1991

[6] Daojing He, Sammy Chan, Shaohua Tang, Chun Chen, Jiajun Bu and Pingxin Zhang 2013,'A Novel and Lightweight System to Secure Wireless Medical Sensor Networks',IEEE Journal of Biomedical and Health Informatics, vol.18,No.1,pp.316-326.

[7] Daojing He, Sammy Chan, Yan Zhang 2014, 'Lightweight and Confidential Data Discovery and Dissemination for Wireless Body Area Networks', IEEE journal of Biomedical and Health Informatics, Vol. 18, No.2, pp.440-448.

[8] Deirdre Morris, Shirley Coyle, Yanzhe Wu, King Tong Lau, Gordon Wallace and Dermot Diamond 2009, 'Biosensing textile-based patch with integrated optical detection system for sweat monitoring', Sensors and Actuators B: Chemical 139, pp. 231–236.Ery

[9] Un-Liu, Jimin Liang, Liaojun Pang, Min Xie and Jie Tian 2010, 'Minutiae and modified Biocode fusion for fingerprint-based key generation', Journal of Network and Computer Applications,

[10] Sriram Cherukuri, Krishna K Venkatasubramanian and Sandeep K S Gupta 2003, 'Biosec: A biometric based approach for securing communication in wireless networks of biosensors implanted in the human body', in Proc. 2003 Int. Conf. Parallel Processing Workshops, pp. 432–439.

[11] Shu-Di Bao, Yuan-Ting Zhang and Lian-Feng Shen 2005, 'Physiological signal-based entity authentication for body area sensor networks and mobile healthcare systems', in 27th Annual International Conference of the IEEE EMBS, pp. 2455–2458

[12] Patrick Traynor, Guohong Cao and Tom La Porta 2006, 'The Effects of Probabilistic Key Management on Secure Routing in Sensor Networks', Wireless Communications and Networking Conference, WCNC, IEEE, pp. 1-13

[13] Sungdae Choi, Seong-Jun Song, Kyomin Sohn, Hyejung Kim, Jooyoung Kim, Jerald Yoo and Hoi-Jun Yoo 2006, 'A low-power startopology body area network controller for periodic data monitoring around and inside the human body', in Proc. 10th IEEE Int. Symp. Wearable Comput., pp. 139–140

[14] Upkar Varshney 2008, 'A framework for supporting emergency messages in wireless patient monitoring', Decision Support Systems vol. 45, pp. 981–996

[15] VeikkoIkonen and Eija Kaasinen 2008, 'Ethical assessment in the design of ambient assisted living', in Assisted Living Systems – Models, Architectures and Engineering Approaches, pp. 1-8.

[16] Vijay Srinivasan, John Stankovic and Kamin Whitehouse 2008, 'Protecting your daily inhome activity information from a wireless snooping attack', in 10th International Conference on Ubiquitous Computing, ACM, pp. 202–211

[17] Young-Dong Lee and Wan-Young Chung 2009, 'Wireless sensor network based wearable smart shirt for ubiquitous health and activity monitoring', Sensors and Actuators B: Chemical, vol. 140, no. 2, pp. 390–395

[18] Eryun Liu, AB, Jimin Liang, A, Liaojun Pang, C, Min Xie, C, JieTiana, DE, Liu, J, Liang, L, Pang, M, Xie and Tian, GH 2010, 'Minutiae and modified biocode fusion for fingerprint-based key generation', J. Netw. Comput.Appl., vol. 33, pp

[19] Krishna K Venkatasubramanian and Sandeep, KS Gupt 2010, 'Physiological value based efficient usable secutity solutions for body sensor networks', ACM Transactions on sensor networks, vol. 6, no. 4

[20] Lifang Wu, Peng Xiao, Songlong Yuan, Siyuan Jiang and Chang Wen Chen 2011, 'A Fuzzy Vault Scheme for Ordered Biometrics', Journal of Communications, vol. 6, no. 9, pp. 682-690

[21] Min Chen, Sergio Gonzalez, Athanasios Vasilakos, Huasong Cao and Victor CM Leung 2011, 'Body area networks: A survey', Mobile. Networks and Applications, vol. 16, no. 2, pp. 171–193

[22] Zhaoyang Zhang, Honggang Wang, Athanasios V Vasilakos and Hua Fang 2012, 'ECG-Cryptography and Authentication in Body Area Networks', IEEE Transactions on Information Technology In Biomedicine, vol. 16, no. 6, pp.1070-1078

[23] Daojing He, Sammy Chan, Shaohua Tang, Chun Chen, Jiajun Bu and Pingxin Zhang 2013,'A Novel and Lightweight System to Secure Wireless Medical Sensor Networks',IEEE Journal of Biomedical and Health Informatics, Vol.18, No.1, pp.316-326

[24] Daojing He, Sammy Chan, Yan Zhang 2014, 'Lightweight and Confidential Data Discovery and Dissemination for Wireless Body Area Networks', IEEE journal of Biomedical and Health Informatics, Vol. 18, No.2, pp.440-448

[25] Najeh Jammali, Lamia Chaari Fourati 2015, 'PFKA: A physiological feature based key agreement for wireless body area network', in Proceedongs of IEEE International Conference on Wireless Networks and Mobile Communications (WINCOM), pp.1-8.

[26] Hussein Moosavi and Francis Minhthang Bui 2016,'Delay-Aware Optimization of Physical Layer Security in Multi-Hop Wireless Body Area Networks',IEEE Transactions on Information Forensics and Security,Vol.11, No.9, pp.1928-1939.

[27] Tilendra Choudhary and M. Sabarimalai Manikandan 2016,'Robust Photoplethysmographic (PPG) Based Biometric Authentication for Wireless Body Area Networks and m-Health Applications', in proceedings of IEEE twenty second National Conference on Communication (NCC), pp.1-6.

8

Health Informatics System using Machine Learning Techniques

Sindhu Rajendran[1], N. Ramavenkateswaran[1], Prashant Abbi[1], Khushi Arora[1], Praveen Kumar Gupta[1], and Mayank Anand[2]

[1]RV College of Engineering, India;
Emails: sindhur@rvce.edu.in; ramavenkateswarann@rvce.edu.
in; prashantabbi.is19@rvce.edu.in; khushiarora.cs19@rvce.edu.in;
praveenkgupta@rvce.edu.in
[2]UAE University, UAE;
Email: gururani@uaeu.ac.ae

Abstract

A global pandemic is the cause of concern for humanity. The data collection and their analytics are a critical part of research and clinical studies for decision-making activities in the healthcare sector. Healthcare informatics systems and analytics (HCI&A) is a rapidly emerging technology in the medical domain that could be explored for analyzing pandemics like coronavirus disease 2019 (COVID-19). The ethical, legal, and privacy issues to be considered during data collection for research activities. Data governance and data stewardship are required to be addressed during interoperability and interpretation while sharing and reusing the data in collaborative research. The sharing of comprehensive records of clinical data collected by EHRs, also known as electronic health records, to be stored and analyzed on a time-to-time basis. The emerging area of information technology, represented by big data and artificial intelligence (AI) technology, has been widely studied in recent circumstances like COVID-19 for pandemic management. The possibility of using machine learning is explored for better predictive diagnostics and treatment.

This chapter discusses the application of artificial intelligence in pandemic management including prevention, diagnosis, treatment, and also critical policy decisions in the COVID-19 pandemic. The methods to collect the digital data of health records are categorized along with few constraints as most of the electronic records related to clinical and epidemiological data are obtained through a shared database such as national and international collaborative informatics infrastructure. The necessity of digital technologies for pandemic emergencies including medical infrastructure reorganization and data workflow model is highlighted. A comparative study of different machine learning models is discussed in the subsequent sections. The digital healthcare informatics envisage a decentralized network architecture and better privacy and security such as blockchain and heterogeneous data collection with machine learning capability are also emphasized.

8.1 Introduction

8.1.1 *COVID-19 pandemic*

COVID-19 is reported to have been detected first in December of 2019 in Wuhan, China. The virus spread quickly since the initial outbreak and spread to various countries all over the world. Even though this virus exhibits origins very close to SARS, or severe acute respiratory syndrome, it is reported to spread much quicker.

Keeping in mind that the first cases were from China, a lot of the research has been focusing on the outbreak there. This includes factors like the biological properties, the risk factors, and the transmission trends of the virus. Lately, however, the research has been covering other regions of the world as well.

A few examples of the same would be the studies that have been conducted in the various regions of Asia such as the outbreak in Japan on the Diamond Princess cruise ship This study used a Bayesian framework along with a Hamiltonian Monte Carlo algorithm [1]. Simultaneously, the ascertainment rate was estimated using a Poisson process in Japan [3]. Another study was conducted in South Korea modeling the evolution and effective reproduction numbers of the virus using Susceptible-Infected-Susceptible models [2]. In India, the basic reproduction number was modeled using a classical Susceptible-Exposed-Infectious-Recovered-type compartmental model [4] and the number of cases were forecasted in the different states using deep learning based models [5].

Even in North America and South America, similar methods have been used for analyses. A few examples for this would include the progression modeling of the outbreak in the US with a Susceptible-Infected-Recovered model until 2021 [6], the prediction of trend of the epidemic in Peru and Brazil using logistic growth model and ML techniques [7], etc. Another study in the US consisted of the spatial variability of the incidence using spatial lag and error models along with geographically weighted regression models [8]. Further, in the United States of America, the estimation of death toll was done using a modified logistic fault-dependent detection model [9] while the infection rates and prevalence across the various states was studied with the help of a sample selection model [10]. In Colombia, the relationship between the incidence and social media communication was investigated using non-linear regression models.

Coming to Africa, the spread of the virus was simulated and predicted in Egypt, Nigeria, Kenya, Algeria, South Africa, and Senegal using a modified Susceptible-Exposed-Infectious-Recovered model [11]. In West Africa, the spread of the virus was predicted using a deterministic Susceptible-Exposed-Infectious-Recovered model [12]. In East Africa, a study was done to forecast the prevalence of the virus using autoregressive integrated moving average models [13]. In Nigeria, ordinary least squares regression was used to predict the spread of the virus using personnel contact and travel history [14]. In Saudi Arabia, generation of forecasting of daily confirmed cases was done using Susceptible-Infected-Recovered models and logistic growth [15].

Apart from the many classical models used as stated above, a lot of the more up-to-date advancements in the statistics literature have led to several new models that could possibly be used in the modeling of infectious diseases. These include dynamic time warping, model selection and combination, and mixed frequency analysis.

COVID-19, being such an infectious disease, needs to be controlled, but doing so is not only time critical but also difficult. The most important factor is the health of the population all over the world since all the research is going towards vaccines and governments are trying to carry out health measures for the public to curtail the spread of the virus. Countries have been implementing measles such as lockdowns wherein people are advised and/or required to remain at home unless absolutely necessary. Nonetheless, the implications of the pandemic have impacted not just the healthcare sector but also sectors such as education, environment, and economy.

Since the case count of individuals infected with the virus kept increasing, the pressure on healthcare providers and medical personnel rose rapidly to test and diagnose the potentially infected individuals besides the usual medical services that are required in general. Invariably, the attempt to curb the virus has led to deprivation and backlog of usual medical procedures [16] since healthcare personnel are still trying to figure out a balance. The restrictions implemented on the movement of people have led to an increase in possibility of increased psychiatric disorders due to the stress and anxiety. These may have long-term negative impact on the mental health of people and might be related to suicidal behaviors and morbidity [17], [18].

Along with restrictions such as the lockdowns, many countries have put a halt on all non-essential businesses that eventually have had to close down permanently causing a substantial increase in unemployment and a heavy negative impact on the economies of the countries. The tourism and travel industries have been severely affected due to the limits on local and international travel leading to a negative impact on the countries heavily dependent on tourism for their income. Amongst the many negative impacts of the pandemic, there are also some positive ones such as businesses adapting newer methods of functioning. A few examples of the same would be the banking industry dealing with higher credit risks while the insurance industry is evolving and adapting newer digital products that deal with solutions more pandemic focused. The automotive industry is expected to deal with a profit reduction of nearly $100 billion that might be due to the software subscription services of modern-day vehicles. Many businesses have shifted to remote work and effectively reduced their costs, and the food industry has drifted towards delivery and takeaway services [19].

Even though the limitations on the businesses that were able to continue operations have been able to lead to possible improvements in the environment mostly from the reduction of pollution [20], the overall societal issues have worsened. The reduction in the labor force due to the restrictions imposed in order to control the spread of the virus has affected women and ethnic minorities the most [19].

Italy and Spain were the first European countries to be massively affected by the virus in Europe. The majority of the literature that is covering the two countries, however, focuses more on the clinical aspects of the infection [21]–[25] while very few explore the prevalence [26]–[28].

8.1.2. Necessity of AI

Artificial intelligence is increasingly becoming more sophisticated, efficient, and quicker at doing what humans do, that too at a lower cost. The prospective that AI and robotics hold in the healthcare sector is vast. Similar to our day-to-day lives, artificial intelligence and robotics are progressively becoming more and more integrated into the healthcare ecosystem. A few use cases are as follows.

Maintaining health

One of the greatest potential benefits of artificial intelligence in the healthcare sector is to help individuals maintain their health so as to not require doctors often. AI and Internet of Medical Things (IoMT) are already helping individuals in customer health applications. These apps encourage proactive management of a healthy lifestyle and keep the consumer in control of their health and well-being.

Furthermore, artificial intelligence enhances the ability of a better understanding of the everyday patterns for the healthcare personnel to assist their patients. This enables them to provide better guidance, feedback, and support.

Early detection

Diseases are being able to be diagnosed more accurately and in the early stages with the assistance of artificial intelligence. An example of such would be cancer. A high proportion of mammograms generates false positives or incorrect results. This leads to healthy women being told they have cancer or worse, diseased individuals not getting diagnosed correctly, etc. Artificial intelligence enables review and translation of mammograms 30% faster with 99% accuracy. This largely reduces the requirement for unnecessary biopsies [29].

The growth of wearable medical devices can also be combined with artificial intelligence. Such devices are being used to overlook heart diseases in its early stages, help medical personnel and other caretakers monitor, and hopefully detect possibly fatal episodes at treatable early stages.

Treatment

Apart from analyzing the risk factor of an adverse episode for a chronically ill patient, artificial intelligence can also help medical personnel take more comprehensive approaches for illness management and help patients manage their long-term treatment programs better.

Machines have been used in the medical ecosystem for more than 30 years. These vary from simple lab machines to severely complicated surgery robots that can either help a human or perform the surgery by itself. Additionally, hospitals and labs use them for repetitive tasks in physical therapy, rehabilitation, and those with long-term comorbidities.

End of life care

Compared to previous generations, we are living longer, and as we near the end of our lives, our death comes in a much slower and different way. More and more illnesses such as heart failure, osteoporosis, and dementia are causing slow deaths. Apart from the illness, even the loneliness plagues this phase of life.

Robots and machines can in all probability revolutionize the care given at the end of life. It can help individuals be independent for longer periods of time and it can help in lowering the requirement of hospitalizations. Artificial intelligence along with the advancements in humanoid robots are allowing robots and machines to go as far as having conversations and other social interactions with humans to keep the minds of the aged agile.

Research

Research in the healthcare field is time consuming and costly. According to the California Biomedical Research Association, an average of 12 years is required for a drug to move from the research laboratories to the point where it is available to the general public. Out of about 5000 drugs that start preclinical trials, only 5 make it to human testing, while just one of these would be approved for human usage. Moreover, it costs a pharmaceutical company around $359 million on an average to develop a new drug, covering bases from the research lab all the way to the patients [30].

One of the more recent applications for artificial intelligence in the healthcare industry is drug research and discovery. The latest advances in AI are directed to streamline the discovery and repurposing of drugs. These processes contain the potential to considerably reduce cost and time required to bring new drugs to the market.

Training

Individuals in training can now go through realistic simulations with the help of artificial intelligence. Such simulations cannot be implemented through basic computer-driven algorithms. The advent of natural speech recognition and AI computers being able to draw on a large database of scenarios almost instantaneously signifies responses to questions, advice, and/or decisions from a trainee can be challenged and checked thoroughly, in a way humans cannot. Due to the self-learning nature of AI, the challenges can be continually adjusted to the trainee's learning requirements based on the previous responses from them.

With advancements in technology, artificial intelligence is now available even on smartphones. Thus, training can be held anywhere with quick catch-up sessions if required after an unusual case or even while traveling.

8.1.3 Artificial intelligence vs. machine learning vs. deep learning

Artificial intelligence, machine learning, and deep learning are components of data science that are, even though very closely related to each other, very different from each other.

Artificial intelligence

Ever since renowned mathematician Alan Turing posed the question, "Can machines think?", humans have been aiming at creating artificial intelligence. Artificial intelligence allows machines to "think". In other words, the machine is able to make decisions without any human intervention. It belongs to the group of computer science that focuses on making machines appear to possess intelligence equivalent to human intelligence.

Artificial intelligence systems can be further classified into various types based on their capability to emulate human behavior, the hardware used to implement so, their real-world applications, and the concept in mind. Based on these factors, all systems of artificial intelligence could fall under one of the following types.

ANI: Artificial Narrow Intelligence

The only subcategory of artificial intelligence that is present in today's world is ANI. It is also known as weak AI. Every task is performed programmatically and is goal oriented. Examples of artificial narrow intelligence would include chat bots, Siri, Alexa, self-driving cars, etc.

Systems that implement ANI are not sentient, conscious, or emotion driven like humans. They utilize data from datasets specified and are unable to perform tasks other than the single task they were designed to perform.

AGI: Artificial General Intelligence

AGI is also known as strong AI and it is more at its conceptual level. In AGI, machines display human intelligence and have the capability to learn, understand, and perform in ways very similar to how humans do in a specific situation. Its behavior is almost indistinguishable from the human's behavior. AGI currently does not exist but has been a part of many science fiction movies wherein machines are not only intelligent but also conscious, self-aware, and driven by emotions.

Strong AI could provide us with the ability to build machines and robots that can think, plan and execute multiple tasks simultaneously, even under uncertain conditions. They would be able to assimilate the prior knowledge they have into the decision-making process to devise innovative, unconventional, and creative solutions.

ASI: Artificial Super Intelligence

ASI is, as of yet, a hypothetical form of artificial intelligence wherein machines would be able to exhibit intelligence that exceeds the capabilities of human intelligence. Apart from having the multifaceted intelligence that humans possess, through this type of artificial intelligence, machines will be able to achieve better decision making and problem-solving capabilities. The hope is that these capabilities will be far superior to that of human beings. However, a possible negative effect of this type of artificial intelligence is that since it would have such a great impact on humanity, it is predicted that if this technology comes into existence, it may lead to the extinction of the human race.

Machine learning

The concept of machine learning uses statistical learning algorithms to create systems that are able to learn and enhance itself from experiences automatically without being explicitly programmed. It is a subset of artificial intelligence.

Services such as recommendation systems like Netflix, YouTube, Spotify, search engines such as Yahoo and Google, and voice assistants such as Alexa and Siri use ML. In machine learning, the algorithms are trained by providing them with tons of data which it learns from by processing the given data.

Machine learning algorithms can be mainly subdivided into three major categories: supervised learning, unsupervised learning, and reinforcement learning.

Supervised learning

In this type of learning, input and output variables are taken and an algorithm is used to learn the mapping between the input and output. Thus, in a supervised learning algorithm, the input would be a known dataset and the output would be its known responses to the input data. Having received this, the algorithm learns the model, be it regression or classification, and then it trains a model to produce a prediction for the output to new data also known as the test datasets.

Unsupervised learning

When the data we have is not labeled, unsupervised learning is used. The main focus of this type of learning is to learn more about the data by deducing patterns in the dataset with no reference to the outputs known. It groups the unsorted data on its own, and thus the name, unsupervised learning. It finds similarities, differences, and patterns in the data to group the data. This type of learning is mostly performed for exploratory data analysis and dimensionality reduction. It is also commonly used to find clusters of data in an unsorted dataset.

Reinforcement learning

Reinforcement learning is the learning by constantly interacting with the environment. In this type of ML algorithm, the agent learns from an interactive environment using a trial-and-error approach by using feedback continuously from its prior actions and experiences. In this type of learning, the agents receive rewards when they perform the right actions and are punished when done incorrectly.

Deep learning

Deep learning is an ML technique that was inspired based on how the human brain filters information, that is, by learning from examples. This helps the model to filter the information through different layers so as to predict and classify the data. Deep learning is mostly applied in applications that people do since it processes information similar to how a human brain does.

Deep learning is the main technology behind self-driving cars since it allows the car to recognize traffic signals and differentiate between a lamp post and a pedestrian. Deep learning methods are also often known as deep neural networks since most of them use neural networks architectures.

8.1.4 Healthcare informatics systems and analytics

Healthcare informatics research has applications in computer technology, information science, statistical modeling techniques, etc., and is a scientific endeavor. It is used to improve patient care outcomes and healthcare provider's performance by developing decision support systems. Strategies involved in the analysis include the following:

- data warehouse formulation for exploration;
- mining of data;

- confirmatory statistical analysis application;
- simulation with the help of an interface with information and computer system technologies;
- translational research.

Healthcare informatics and analytics, also known as HCI&A, has evolved quite fast since the emergence of advanced data analytics technologies being applied in healthcare sector. HCI&A currently is a crucial area of study for both academic researchers and practitioners. Consequently, this emerging field has prompted a plethora of opportunities and challenges. These are in relation to healthcare data management and the application of advanced analytics in the healthcare sector.

8.2 Concept of Blockchain

Blockchain is a method of documenting information in such a way that it is nearly impossible to edit, hack, or deceive the system.

Essentially, a blockchain is a digital ledger of transactions which is replicated and redistributed across the whole network of systems present on the blockchain which uses immutable cryptographic signature. Every block in the blockchain has various transactions and each time a transaction happens, an entry of that particular transaction is appended to each member's ledger. Distributed ledger technology, also known as DTL, is the decentralized database that multiple participants manage.

In the conventional methods, participants that wanted to record transactions and track their assets had to keep their own ledger. This method proved to be costly, partly since it involved intermediaries that charged service and convenience fees. It is also not very efficient considering the delays that took place in the execution of the agreements. The maintenance of multiple ledgers also cost a ton of effort. Another key point to be noted is that in the conventional methods, it is vulnerable since if the central system itself is corrupted due to any reason for example hacks, frauds, or even a simple human error, the entire system and even the whole business would be impacted. Thus, a lot of business networks have now shifted to using blockchain technology. This architecture provides the members with the capability to share a ledger. This ledger is updated each time a transaction takes place using a peer-to-peer replication. This means that each node, also known as the participant, in the network would behave as both the publisher as well as the subscriber.

There are multiple crucial steps involved in any transaction before it is added to the blockchain. A few of them are authentication (with the help of cryptographic keys), authorization (with the help of proof of work), etc.

Authentication

The initial design of blockchain was to be able to operate even without any central authority. The transactions, though, were still to be authenticated. Blockchain achieves this by using cryptographic keys. Cryptographic keys are strings of data that identify a participant and provide access to their "wallet" or "account" on the system. A unique set of private and public keys are possessed by every participant. The public key is visible to everyone. Together, these two keys create a secure and safe digital identity to authenticate the participant using digital signatures so as to "unlock" the transaction that they intend to perform.

Authorization

After the transaction has been authenticated and agreed upon by the participants, it needs to be authorized or approved prior to it being appended to the blockchain.

The finalization of adding a new transaction to a block in the chain, for public blockchains, is done using consensus. In other words, for the transaction to be added, the majority of the computers on the network, also known as nodes, need to agree that it is a valid transaction. The verification of transactions done by individuals that own the computers on the network are provided incentives through rewards. This is referred to as "proof of work".

Proof of work

The process of requiring the individuals that own the computers on the network to solve mathematical problems that are complex in nature so as to be able to add new blocks to the chain is known as proof of work. The solving of such problems is called "mining" and the people that solve such problems are known as "miners". These miners are rewarded usually using cryptocurrency for their work.

As easy as it seems, mining is a difficult process. The problems set have the odds of about 1 in 5.9 trillion of being solved and can be solved only by using trial-and-error methods. This requires a significant amount of computing power that uses a substantial amount of energy. In other words, the rewards for mining must exceed the expenses of the computers as well as the expenses of electricity for running the computers considering that each computer would take years' worth of time to find a solution to the problem.

The issue with proof of work

To increase chances of creating economies of a good scale, miners quite often pool together their resources with the help of companies that assemble large groups of miners that then share the rewards offered by the blockchain.

With more and more computers joining the network to try to solve the problem, the blockchain grows and so does the complexity of the problem. Theoretically, the chain is redistributed further which then makes it harder to hack or sabotage. Practically though, with the concentration of the mining power in just a few mining pools, the organizations are now large enough with a lot of computing power and electrical power to maintain as well as grow the blockchain network that has been implemented around validation via the proof of work.

Proof of stake

Newer blockchain networks have implemented a new type of validation protocol known as "proof of stake". Here, the participants have to have a stake in the blockchain to be eligible for a chance at selecting, verifying, and validating transactions. This stake is usually owning cryptocurrency. Since no mining is required here, a substantial amount of computing power is saved.

Additionally, newer technologies in blockchain have started including "smart contracts". These automatically execute the transactions as soon as certain conditions are met.

In blockchains, the identifier of the last block is included into the identifier in the next block. This creates an immutable and unbreakable chain. The key to keeping the data in the blockchain manageable and secure as more and more blocks are added is through hashing along with Merkle tree which is a consolidating data structure.

Similar to a password generator, after a transaction has been verified, it will be passed through a hash algorithm. This hash algorithm converts it into a tuple of unique letters and numbers. Further, a couple of transaction hashes will be merged and passed through the hash algorithm once more so as to produce a new distinctive hash. This process is continued until there is only one hash remaining. This hash is known as the "root" of several transactions.

Hashes are used as a key security feature for blockchains since they are unique and they work in only one direction. The same data will produce the same hash of letters and numbers every time making hashing consistent and reliable. On the other hand, it is not possible to decipher the original data using the letters and numbers. In other words, it is not possible to reverse the hashing process.

Since the hashing process will create the same hashes if the process is repeated with the same transaction, anyone using the blockchain can check whether the data has been tampered with or not considering any tampering with the data would lead to the generation of a completely different hash. This would then affect all iterations of the hashes up till the root. This is known as a Merkle tree.

The purpose of Merkle trees is to reduce the quantity of data required that is required to be stored and transmitted or be broadcasted through the entire network significantly by summarizing the transactions that have been hashed multiple times into just one root hash. The root hash is of standard size itself since each transaction is hashed, combined with another hashed transaction and hashed again.

8.2.1 Network architecture of blockchain

Blockchain gets its name by its functionality. It consists of blocks that store data which are then linked to one another to form a chain. The size of the blockchain keeps increasing as the number of transactions increase. The architectural components in a blockchain network include:

- Transaction
- Block
- P2P network
- Consensus algorithm

Transaction

A blockchain system's simplest building blocks are transactions. A sender address, a recipient address, and a value are usually included. It resembles a typical credit card statement. The value is transferred by the owner digitally signing the hash created by adding the previous transaction and the receiver's public key.

The transaction is then publicly disclosed to the network, and each node holds its own copy of the blockchain. The current known "state" is calculated by processing each transaction in the blockchain in the order it appears. Transactions are grouped and transmitted in the form of a block to each node. Each node independently verifies and "processes" new transactions as they are propagated over the network. Each transaction is timestamped and grouped together into a block.

Block

A block comprises information in the form of a block header as well as transactions. Blocks are data structures that are replicated to all nodes in the network and are used to bundle collections of transactions. Miners are the ones who create blocks in the blockchain. The process of creating a legitimate block that will be accepted by the rest of the network is known as mining. Nodes package pending transactions into blocks to be stored on the blockchain after verifying that they are cryptographically correct. The metadata in a block header is used to verify the block's validity. Transactions make up the rest of a block. Based on the miner's preference, a block can include any number of transactions.

Types of blocks

The majority of blocks merely add to the current primary blockchain, which is also the network's longest chain. These are known as "main branch blocks". Some blocks point to a parent block that is not on the longest blockchain. "Side branch blocks" are the name for these blocks. Some blocks relate to a parent block that the node executing the block is unaware of. These are referred to as "orphan blocks."

Although side branch blocks are not yet a part of the main branch, if further blocks are mined that identify them as a parent, a specific side branch may be reorganized into the main branch. This introduces the idea of forking.

P2P network

The blockchain is a peer-to-peer (P2P) network that uses the IP to communicate. A P2P network has no centralized nodes and has a flat topology. While collaborating via a consensus process, all nodes can equally give and consume services. Peers contribute to the computational power and storage needed to keep the network running. P2P networks are often more secure than centralized networks because they do not have a single point of attack or failure. Permission-based and permission-less networks are both possible in a blockchain network. Because anybody can join a permission-less network, it is also known as a public blockchain, whereas a permission-based blockchain is known as a consortium blockchain. A permission-based blockchain, sometimes known as a private blockchain, necessitates pre-verification of the network's participants, who are typically known to one another. Every node in a network maintains a local copy of blockchain in a typical blockchain design. The P2P network on which blockchain architecture is based is solely responsible for its decentralization.

Consensus algorithm

A consensus technique is used to synchronize all of these copies of a single ledger. The consensus method assures that whatever local copies each participant has are consistent with one another and are the most up-to-date. Every individual node has copies that are identical or similar to one another. The consensus algorithm might be said to be at the heart of every blockchain design.

SBFT – Simplified Byzantine Fault Tolerance

The main concept here is that a single validator bundles the proposed transactions and creates a new block. Given the permission-based architecture of the ledger, the validator is a known party. The consensus is reached when a small number of other nodes in the network correct the new block.

8.3 Concept of Data Leak and How It Is Overcome using Blockchain

A great amount of fear is elicited in the minds of people in today's world by terms like "hack" and "data breach". Hackers pose tremendous threat to any organization, be it small or large in scale.

Companies have been drafting policies and creating better technologies to protect their data and information since the introduction of the GDPR, also known as the General Data Protection Regulation, and other such similar data security legislations. Amidst these efforts, blockchain technology has been in high consideration that might help enable organizations to safely store sensitive data.

Blockchain is a decentralized technology. This means that the system needs all the participants involved to consent to any new transaction being added to the network. Additionally, transactions once added to a block in the chain cannot be altered since blockchain is a shared ledger.

Blockchain is a self-auditing system. Its network stays in a state of consensus and automatically checks in with itself every 10 minutes. This provides many benefits including the records being essentially public and verifiable since the blockchain is not stored in any particular single location. This allows for faster and safer transactions due to the transparency across all participants involved. Additionally, the blockchain cannot be controlled by a single individual or entity since there is no centralized version of the data in the system, thus making it arduous for anyone to corrupt or hack the data.

In a centralized system, once a hacker has hold of even one version of the data, the rest of the system is essentially in their hands. This is not the case with blockchain since even if the hacker has hold of one of the blocks in the chain, they would have to get access to all the next blocks in order to get hold of the entire system. Thus, if an organization has multiple transactions in the order of, say, millions made each day in the blockchain, it would be nearly impossible to gain access to all the blocks of information in the system. Based on the type of data involved, the consequences of a hack could range from destruction or corruption of data to leaking of confidential information or even intellectual property theft.

8.3.1 Data breach targets

A business becomes a data breach target when their data is of use or of value to a third party. Different types of data will be of varying degree of value based on the varying levels of risk to the said business. A few types of data that might be targeted as are follows:

- Contact information, social security numbers, education, birth dates, and other personal information are examples of personally identifiable information.
- Pre-existing health issues, prescription drugs, therapies, and medical records are all examples of health information.
- Financial data comprises credit/debit card numbers and expiration dates, bank accounts, investment information, and so on.
- Lists of usernames and passcodes, encryption keys, network layout, and security strategies are all examples of IT security data.
- Product drawings and manuals, specifications, marketing texts and symbols, scientific formulas, proprietary software, and other content generated by the company are examples of intellectual property.
- Legal information – This category includes information such as court documents, merger and acquisition details, legal views on corporate operations, and regulatory judgments.
- Data about competitors, pricing information, market studies, and company plans are all examples of competition information.

The above types of information are of value to third parties and attract their attention. Financial-, personal-, and health-related data that can be used for fraud, identity theft, or marketing can be easily sold for monetary benefit. The sale of IP-related data could be misused by competitors to hinder the company's plans while legal data's leak could harm the company's legal position. Similarly, IT-related data would be a very valuable target since unauthorized personnel would get access to all the other data stored on the system.

In the healthcare sector, information regarding any patient's health records is of utmost importance and should be considered sensitive data since it is confidential. If this data is hacked or corrupted, it can pose a major threat to patients in terms of the past health records being tampered with. Essentially, this would affect the decisions made by the medical personnel and could, in worst-case scenarios, prove to be fatal. Incorporation of blockchain would reduce the chances of such mishaps tremendously and safeguard the general public health as a whole.

8.3.2 Data breach threats

Threats with respect to data breach can be initiated by employees of the company, suppliers, consultants, or any individual that has access to the network or even from outside the company. A few ways access to the data can be gained through external email accounts, through the cloud, or through mobile devices. Traditional protection systems are not enough anymore to keep data safe.

Against insiders, data protection techniques could fail since unhappy employees might decide to leak information for personal benefit. People outside the organization could access emails or release malware into employees' computers to access their credentials. In wake of such threats, businesses now have to come up with consequences of such data breaches to find solutions and decrease risks.

8.3.3 Data breach consequences

Businesses are increasingly experiencing severe consequences due to data breaches. This is majorly due to the increase in regulatory burden for notification of the persons whose data has been compromised. Based on jurisdiction, notification requirements and penalties differ for businesses suffering a data breach.

A company whose data has been breached needs to establish where their customers live and the regulatory authorities that have jurisdiction over the said area. These authorities define the regulations for notifications based on the type of data that has been compromised. This includes who the notification needs to be sent to, how the notification needs to be sent, and whether or not certain authorities need to be notified. Usually, breaches involving financial, personal, and health data are subject to these requirements. Businesses with an international customer base might have customers residing in multiple jurisdictions, and the requirements, thus, would vary. The cost of such compliance along with legal penalties, potential compensation for harm, and possible lawsuits are extremely high and can cause serious monetary losses for the business.

Since blockchain technology is based on encryption and is spread across a vast network of read-only computers, it helps in the maintenance of a safer record and adds security similar to an impenetrable wall. Blockchain is based on complete accountability. The system can be fully audited by an end user and they can verify that it is operating exactly as advertised. Companies now are increasingly embracing public blockchain technology as a competitive advantage.

8.4 Present Situation and Future Perspective

The healthcare industry is complicated and intricate. Even though this sector has gone through profound innovation in aspects such as vaccines, drugs, medicines, clinical trials, and even the adoption of various new technologies like cloud computing, it still requires progress in aspects such as patient data, data analysis, supply chain management, drug traceability, cybersecurity, etc.

Blockchain has the capacity of eliminating the issues mentioned above and redefining the healthcare sector by putting the patient at the core of the healthcare ecosystem. It can also establish increased security, interoperability, and privacy and safety of the stored patient records. The value of the blockchain technology was $2.12 billion in the healthcare industry in 2019 and with a CAGR of 8.7% throughout the forecast period (2020–2025) is estimated to reach $3.49 billion by the year 2025.

Thus, increasingly, the scope of blockchain technology in the healthcare industry for patient records is being explored by various companies. From enhancing payment options and tracking pharmaceuticals to decentralization of patient data, blockchain can prove to be an irreplaceable tool.

A few ways blockchain is revolutionizing the healthcare industry is as follows.

8.4.1 Drug traceability

For decades now, drug counterfeiting has been a major problem for the healthcare sector. According to the Organization of Health Research Funding:

- 10%–30% of medications launched into developing-country markets are counterfeit or copied versions.
- Online sales of these cloned drugs annually accounts for $80 billion.
- The annual market for counterfeit pharmaceuticals is over $200 billion.
- In 2014, around 60 different types of medicines and products being produced, the renowned pharmaceutical company Pfizer was counterfeited. According to the WHO, 16% of counterfeit pharmaceuticals consisted of wrong chemicals or included the incorrect amount of substances, endangering the patients and thus the entire healthcare sector.

Such counterfeit medicines are substantially different from the original medicine in both composition and quality. Thus, they can prove to be highly dangerous to patients since the product does not have the ingredients it claims to have. In other words, the "medicine" does not treat the disease and might, on the other hand, cause further harm to the patient due to unexpected side effects.

Since blockchain is known for its security and immutability, all records, be it old or new, are time stamped to record it. These time stamped records are immutable. This guarantees that the recorded data is sacrosanct. The technology does not allow an individual to alter the data and thus helps in tracking of any product.

One of the biggest problems that stakeholders in the healthcare sector confront is that they have an incomplete perspective of the supply map, rather than the conformity of the origin of the products that are released on the market. As a result, counterfeit pharmaceuticals enter the supply chain and reach patients. Stakeholders would have access to all recorded information and transaction history via blockchain, allowing them to keep track of every product in the supply chain that is delivered to the market.

8.4.2 Clinical trials

A significant amount of patient data and health information is generated before and after the phases of a clinical trial. Blood test results, statistics, quality reports, health surveys, and clinical documentation from a large number of patients are included. Because the process is so demanding and protracted, it is difficult to keep track of everyone. Unintentionally and frequently intentionally, mistakes are made. Because pharmaceutical companies and sponsors fund these trials with millions of dollars, the financial consequences can be severe, depending on the outcome. If the trial results are positive and lead to the development of a new vaccine or treatment, there will be a large profit or a large financial loss as a result. Fraudulent behavior is frequently used to avoid the latter. This begins with the manipulation, alteration, or concealment of critical data that could jeopardize the clinical trial's progress. As a result, controversies occur, compromising the organization's credibility among authoritative bodies and regulatory patients.

When you use a keyword to search the network, blockchain will display the search results, which will include all the relevant records or documents. If any new information needs to be added to the ledger using the transactional model, the nodes of the ledger must agree that it is reasonable and valid in light of the data's history. If any network participant wanted to change an existing stack of health data, they would have to change it on all of the network machines. This crucial aspect of blockchain technology prevents data from being tampered with in clinical trials. Every piece of trial data is given a unique code that corresponds to the document's content. If a healthcare practitioner examines the recorded data and has doubts about its legitimacy, they may easily confirm it by comparing it to the original data saved in the blockchain system. To accomplish so, they must execute the data, which will result in a new code. That new code must be compared to the original. If they are the same, it means the saved health information has not been tampered with in any way. This will increase collaboration between healthcare professionals and third parties such as pharmacies, paving the way for additional scientific advancements, particularly in the treatment of uncommon diseases.

8.4.3 Management of patient data

There are now two major difficulties that healthcare professionals dealing with patient data management are dealing with. To begin with, patients have their own variability. A cure that is provided to one patient may not be prescribed to a different patient since it is not appropriate for them. Access to their medical records is essential for this reason, in order to generate individualized treatments that are centered around the patient.

Second, keeping such information and providing access to it, as well as sharing it among medical leaders, is a significant barrier. When it comes to transferring highly sensitive personal medical data, a secure network is the way to go. Scientific breakthroughs necessitate the storage of medical data in secure cybersecurity structures, and the lack of such structures is a major roadblock to their growth. Furthermore, the lack of a single database causes this information to be stored in bits and pieces across multiple sites, posing a significant difficulty for researchers. Numerous sites need to be searched in order to reference the required information which slows down the process even further.

There is also the matter of data ownership to consider. A patient cannot have complete control over his or her medical records. They have the ability to change or delete whatever information they want. This could have major health consequences, putting the patient's safety in jeopardy. However, the disadvantage is that if the patient is not authorized to own the data, they will have no idea who is accessing and using it.

Blockchain is based on well-established cryptographic techniques, providing an appropriate cybersecurity structure for data sharing. Healthcare providers enter the patient's name, date of birth, prescription, procedures performed, and ambulatory records into an electronic health record (EHR). This data is stored in cloud computing or existing databases. Within the blockchain system, all of these data sources are tied to a specific hash with the patient's assigned public ID. Through the use of smart contracts, the patient controls data access.

Stakeholders can use an API to search for data, find out where it is, and receive it. An API will provide the patient's customized health information but will not divulge anything else, such as the patient's identity. Smart contracts can be used by a patient to show their whole medical history to any stakeholder. Medical records' access in smart contracts is provided to patients using an API.

The medical business relies extensively on blockchain, as well as other advanced technology such as machine learning and artificial intelligence. Some real-world examples of how blockchain is transforming the healthcare industry through collaborations are as follows.

Transformation of the Drug Supply Chain

According to a market intelligence analysis published by a reputable research organization, blockchain technology has the potential to improve the security, privacy, data provenance, and functionality of the pharmaceutical supply chain if properly applied.

The medical supply chain has been looking into solutions to fight the problem of drug counterfeiting and falsification. Viant, a blockchain-based company, is giving supply chain solutions to pharma giants like GlaxoSmithKline in order to entirely eliminate the problem. The program is built on blockchain tracking technology, which will fine-tune the medical industry's supply chain.

Bookkeeping of clinical trials

Clinical trials have become the new normal in the current situation. Using the bookkeeping process to maintain accurate track of their finances is a critical demand at the moment. It is required for the smooth operation and evaluation of clinical trials. This is where blockchain companies have come up with new ways to make bookkeeping and financial reporting more efficient. Boehringer Ingelheim has teamed up with IBM to develop a blockchain-based bookkeeping solution. The database is intended to highlight their very first clinical trial project. The many parts of the bookkeeping process will be improved tenfold as a result of this.

Process of form filling

When someone needs to be admitted to the hospital for a serious disease, the procedure of admission takes a long time due to the tedious process of filling out numerous documents. It appears to be as exhausting as climbing Mt. Everest! Companies like Kalibrate Blockchain are using blockchain to automate tasks, saving us time and energy. It has developed FormDrop, a mobile application based on the Universal Patient Index blockchain-powered engine. Anyone can use this software to fill out paperwork ahead of time before going to a healthcare establishment. They can save time by not having to wait in lines this way. This will help to boost the productivity of healthcare workers.

The medical healthcare blockchain's benefits

The following benefits and advantages can be inferred from the use of blockchain technology based on the practical applications of blockchain and the healthcare industry difficulties that it can solve.

Robust monitoring

Powerful and efficient monitoring of transactions is primarily one of the most important things that the healthcare network management team requires right now. The aspect that distinguishes blockchain from other healthcare management systems is that it creates a decentralized record of all transactions. It is exact and transparent to the core, saving time, effort, and money, as well as the trouble of ongoing management.

Collaboration enhancement

Any healthcare initiative's success is heavily reliant on the people involved, such as third parties, experts, and researchers. With the use of distributed ledger technology, blockchain technology improves collaboration among these individuals, allowing them to contribute deep insights and conduct group research.

Data protection

The fundamental concern that the healthcare business is facing, as discussed in the preceding sections, is the leaking of significant data and its use for harmful purposes and other vested interests. Another priority area is ensuring that database participants and parties have access to the most up-to-date and accurate patient data and diagnosis information. Data over a blockchain cannot be tampered easily or even changed for that matter due to multiple layers of cryptography components, thus addressing the major problems involving the breach of security.

Cure cost-effectiveness and process simplicity

Robust, efficient, and superior data sharing capabilities coupled with rapid distribution among the stakeholders in the healthcare industry improve the development of budgeted treatments and cures for a variety of diseases.

The detailed procedure records transactions in the form of EHRs (electronic health records). These are the only forms in which the records are digitally available. Simplification of complex operations and increased security by removing additional validations is a direct result of the same.

Medical blockchain predictions for the future

The potential of blockchain in healthcare appears to be extremely promising, since it aids in the resolution of some of the industry's most urgent difficulties. We anticipate a progressive future since blockchain is decentralized, unlike the majority of healthcare records, which are centralized. A future in which blockchain is used as part of a system in which patients manage their own medical data rather than depending on a central source. Of course, the use of blockchain has its own set of technical difficulties. The process of recording, finding, and retrieving data can be complicated and time-consuming, but with the help of upcoming technologies like ML and artificial intelligence, these issues will soon be handled.

8.5 Existing Challenges in the Future

Blockchain is a new technology that is gaining traction in a variety of industries [41] and offers a lot of benefits [42] and potential [43]. This technology, however, presents with issues that must be handled. In this section, we will look at a few of the most pressing issues.

8.5.1 Data security and privacy

The privacy and security of data [44] is the foremost important challenge. The necessity for a third party to make a transaction is removed with the introduction of blockchain-based applications [42]. The data becomes vulnerable to potential privacy and security problems since the blockchain technology allows the entire community, rather than a single trusted third party, to verify the records in a blockchain architecture [45]. The data privacy will be compromised because all nodes have access to the data that has been transmitted by a single node.

When a third party is not present for authorization, the patient must designate one or more than one representative that has access to their medical history and information on their behalf in an emergency. This representative can now grant access to a group of people to the same patient's records, posing a significant data privacy and security risk. When high-security techniques are applied to data, it creates barriers in transmitting data from one block to another, leaving recipients with restricted or missing information. Furthermore, blockchain networks are vulnerable to what is known as a 51% assault [46]. This attack is carried out by a group of miners who control more than half of the blocks in a blockchain network. The miners gain network authority and, if they are not given consent, they can block any new transactions from taking place. Five cryptocurrencies have lately been targeted by this hack, according to coindesk [47]. Furthermore, a patient record may contain sensitive information that should not be stored on the blockchain [48].

8.5.2 Storage capacity management

A new issue that arises in this regard is storage capacity management. Blockchain does not need humongous storage as it was built to record and execute data involving transactions [49]. Gradually, as it expanded into the realm of healthcare, the storage issues became apparent. On a daily basis, the healthcare industry must process a vast amount of data. In the blockchain scenario, everything from patient records, health histories, and test reports to X-rays, MRI scan reports, and other medical imagery will be available to all nodes in the chain, necessitating a vast storage space [ax, ay]. Furthermore, because blockchain applications are transaction-based, the databases utilized to support this technology have a proclivity to develop rapidly. As databases grow in size, the speed at which records can be searched and accessed decreases, making them unsuited for transactions wherein speed is critical. As a result, a blockchain system must be scalable and durable [52].

8.5.3 Interoperability issues

Interoperability [44] is another challenge with blockchain, which involves making blockchains from diverse communication services and providers talk to one another in a seamless and suitable manner. This problem makes it difficult to share data effectively [53].

8.5.4 Standardization challenges

Since blockchain technology is still in its initial stages, it will undoubtedly confront standardization issues as it moves closer to practical adoption in medicine and healthcare. International standardization authorities would be required to provide a number of well-authenticated and approved standards. These predefined standards could be useful in determining the nature, size, as well as the format of data transmitted in blockchain applications. These standards will not only evaluate the shared data, but they will also function as safety precautions.

8.5.5 Social challenges

Because blockchain technology is still in its early stages, it confronts societal problems such as cultural shifts in addition to the technological challenges stated above. Even though the healthcare sector is gradually moving toward digitization, there is still a long road ahead of us before it fully embraces this technology, particularly blockchain, which has yet to be clinically verified.

It will take time and effort to persuade doctors to abandon paper in favor of technology. The technologies and strategies given are somewhat untrustworthy due to their poor adoption rate in the health industry [52]. We cannot, at this time, describe it as a feasible and universal answer for all healthcare difficulties [42] because of all of these challenges and risks.

Conclusion

Blockchain will be critical in transforming the healthcare system and ensuring the future of healthcare. With medical blockchain storing and safeguarding vast volumes of data at scale, the notion of offering tailored treatment no longer seems like an unachievable dream. By the end of next year, we hope that blockchain-based production applications will be mainstream.

Despite the fact that healthcare sometimes lags behind other industries in terms of obtaining new breakthroughs, due to the integrity of the prospective benefits, now may be an appropriate time to embrace blockchain in the healthcare industry. There could be a few main points when it comes to managing electronic health records on the blockchain. Smart contracts would create an indestructible chain of blocks that would take into account individualized care while remaining within clinical frameworks. Converging clinical systems would use a blockchain smart contract to prevent any duplicate copies from being created in the parent centralized system. Researchers may benefit from blockchain since it can give verifiable and timestamped versions of scientific investigation, thereby allowing researchers to have a long-term history of their discoveries. Blockchain is a must-have technology in the massive pharmaceutical sector.

It is possible to decrypt all pharmaceutical association report distribution using blocking. This kind of approach will increase the rapidity with which data is prepared, ensure the accuracy of records distribution, and limit the risk of archive loss, damage, or fabrication. The innovation itself is in charge of these things of interest: The created block cannot be altered or deleted. The blockchain will ensure that data cannot be tampered with. Any offer of substandard or counterfeit drugs will be turned down. This is accomplished by maintaining control over all elements of the pharmaceutical supply chain, including production, logistics, and distribution.

Medical practitioners, R&D specialists, healthcare entities, healthcare providers, and biomedical researchers will benefit from the practical application of blockchain technology, since it will allow them to more effectively distribute large amounts of information, share clinical knowledge, and convey recommendations while maintaining greater privacy and security protection. A blockchain might be used to store a digitized brain, which could be employed in neural-control systems. Only a few organizations have confirmed that blockchain technology can play a role in neurotechnology, which is still in its early phases.

However, it is unclear how secure personal brain data will be stored on a blockchain. Although the open structure of blockchains will undoubtedly prevent unauthorized modifications and tampering of data, many of the usual worries about data collection on a large-scale remain: third-party involvement in transaction of sensitive data for dubious marketing objectives; individuals who could still be identified indirectly through pseudonymous identifiers or data patterns. As a result, this blockchain-based healthcare framework will encourage individuals to participate more actively in their healthcare, as a result improving the quality of life in a more appropriate manner.

References

[1] Mizumoto K., Kagaya K., Zarebski A. and Chowell G., 2020. Estimating the asymptomatic proportion of coronavirus disease 2019 (COVID-19) cases on board the Diamond Princess cruise ship, Yokohama, Japan, 2020. Eurosurveillance, 25, 2000180. Pmid:32183930

[2] Park H. and Kim S.H., 2020. A Study on Herd Immunity of COVID-19 in South Korea: Using a Stochastic Economic-Epidemiological Model. Environmental and Resource Economics, 76, pp. 665–670.

[3] Omori R., Mizumoto K. and Nishiura H., 2020. Ascertainment rate of novel coronavirus disease (COVID-19) in Japan. International Journal of Infectious Diseases, 96, pp. 673–675. Pmid:32389846

[4] Sarkar K., Khajanchi S. and Nieto J.J., 2020. Modeling and forecasting the COVID-19 pandemic in India. Chaos, Solitons & Fractals, 139, 110049.

[5] Arora P., Kumar H. and Panigrahi B.K., 2020. Prediction and analysis of COVID-19 positive cases using deep learning models: A descriptive case study of India. Chaos, Solitons & Fractals, 139, 110017. Pmid:32572310

[6] Atkeson, A., 2020. What Will Be the Economic Impact of COVID-19 in the US? Rough Estimates of Disease Scenarios. National Bureau of Economic Research, Working Paper 26867.

[7] Wang P., Zheng X., Li J. and Zhu B., 2020. Prediction of epidemic trends in COVID-19 with logistic model and machine learning technics. Chaos, Solitons & Fractals, 139, 110058. Pmid:32834611

[8] Mollalo A., Vahedi B. and Rivera K.M., 2020. GIS-based spatial modeling of COVID-19 incidence rate in the continental United States. Science of The Total Environment, 728, 138884. Pmid:32335404

[9] Pham H., 2020. On Estimating the Number of Deaths Related to Covid-19. Mathematics, 8, 655.

[10] Benatia, D., Godefroy, R. and Lewis, J., 2020. Estimating COVID-19 Prevalence in the United States: A Sample Selection Model Approach.

[11] Zhao Z., Li X., Liu F., Zhu G., Ma C. and Wang L., 2020b. Prediction of the COVID-19 spread in African countries and implications for prevention and control: A case study in South Africa, Egypt, Algeria, Nigeria, Senegal and Kenya. Science of The Total Environment, 729, 138959.

[12] Taboe H.B., Salako K.V., Tison J.M., Ngonghala C.N. and Kakaï R.G., 2020. Predicting COVID-19 spread in the face of control measures in West Africa. Mathematical Biosciences, 328, 108431. Pmid:32738248

[13] Takele R., 2020. Stochastic modelling for predicting COVID-19 prevalence in East Africa Countries. Infectious Disease Modelling, 5, pp. 598–607. pmid:32838091

[14] Ogundokun R.O., Lukman A.F., Kibria G.B.M., Awotunde J.B. and Aladeitan B.B., 2020. Predictive modelling of COVID-19 confirmed cases in Nigeria. Infectious Disease Modelling, 5, pp. 543–548. pmid:32835145

[15] Alboaneen D., Pranggono B., Alshammari D., Alqahtani N. and Alyaffer R., 2020. Predicting the Epidemiological Outbreak of the Coronavirus Disease 2019 (COVID-19) in Saudi Arabia. International Journal of Environmental Research and Public Health, 17, 4568. pmid:32630363

[16] Søreide K., Hallet J., Matthews J.B., Schnitzbauer A.A., Line P.D., Lai P.B.S., et al. 2020. Immediate and long-term impact of the COVID-19 pandemic on delivery of surgical services. British Journal of Surgery, 107, pp. 1250–1261. Pmid:32350857

[17] Cao W., Fang Z., Hou G., Han M., Xu X., Dong J, et al. 2020. The psychological impact of the COVID-19 epidemic on college students in China. Psychiatry Research, 287, 112934. pmid:32229390

[18] Torales J., O'Higgins M., Castaldelli-Maia J.M. and Ventriglio A., 2020. The outbreak of COVID-19 coronavirus and its impact on global mental health. International Journal of Social Psychiatry, 66, pp. 317–320. Pmid:32233719

[19] McKinsey & Company, 2020. COVID-19: Implications for business

[20] Forster P.M., Forster H.I., Evans M.J., Gidden M.J., Jones C.D., Keller C.A., et al. 2020. Nature Climate Change, 10, pp. 913–919.

[21] Palmieri L., Vanacore N., Donfrancesco C., Noce C.N., Canevelli M., Punzo O., et al. 2020. Clinical Characteristics of Hospitalized

Individuals Dying With COVID-19 by Age Group in Italy. The Journals of Gerontology: Series A, 75, pp. 1796–1800.

[22] Galván Casas C., Catalá A., Carretero Hernàndez G., RodríguezJiménez P., Fernández-Nieto D., Rodríguez-Villa Lario A., et al. 2020. Classification of the cutaneous manifestations of COVID-19: a rapid prospective nationwide consensus study in Spain with 375 cases. British Journal of Dermatology, 183, pp. 71–77. pmid:32348545

[23] Berenguer J., Ryan P., Rodríguez-Baño J., Jarrín I., Carratalà J., Pachón J., et al. 2020. Characteristics and predictors of death among 4035 consecutively hospitalized patients with COVID-19 in Spain. Clinical Microbiology and Infection, In Press. pmid:32758659

[24] Caruso D., Zerunian M., Polici M., Pucciarelli F., Polidori T., Rucci C., et al., 2020. Chest CT Features of COVID-19 in Rome, Italy. Radiology, 296. pmid:32243238

[25] Lodigiani C., Iapichino G., Carenzo L., Cecconi M., Ferrazzi P., Sebastian T., et al. 2020. Venous and arterial thromboembolic complications in COVID-19 patients admitted to an academic hospital in Milan, Italy. Thrombosis Research, 191, pp. 9–14. Pmid:32353746

[26] Giordano G., Blanchini F., Bruno R., Colaneri P., Di Filippo A., Di Matteo A, et al. 2020. Modelling the COVID-19 epidemic and implementation of population-wide interventions in Italy. Nature Medicine, 26, pp. 855–860. pmid:32322102

[27] Ceylan Z., 2020. Estimation of COVID-19 prevalence in Italy, Spain, and France. Science of The Total Environment, 729, 138817. pmid:32360907

[28] Yuan J., Li M., Lv G. and Lu K., 2020. Monitoring transmissibility and mortality of COVID-19 in Europe. International Journal of Infectious Diseases, 95, pp. 311–315. pmid:32234343

[29] http://www.wired.co.uk/article/cancer-risk-ai-mammograms

[30] California Biomedical Research Association. New Drug Development Process. http://www.ca-biomed.org/pdf/media-kit/fact-sheets/CBRADrug Develop.pdf

[31] http://beamandrew.github.io/deeplearning/2017/02/23/deep_learning_101_part1.html

[32] https://contentsimplicity.com/what-is-deep-learning-and-how-does-it-work/

[33] https://towardsdatascience.com/what-are-the-types-of-machine-learning-e2b9e5d1756f

[34] https://codebots.com/artificial-intelligence/the-3-types-of-ai-is-the-third- even-possible

[35] https://www.sciencedirect.com/topics/computer-science/super vised-learning

[36] https://medium.com/@tjajal/distinguishing-between-narrow-ai-general-ai-and-super-ai-a4bc44172e22

[37] https://towardsdatascience.com/understanding-neural-networks-from-neuron-to-rnn-cnn-and-deep-learning-cd88e90e0a90

[38] https://en.wikipedia.org/wiki/Recursive_neural_network

[39] https://www.youtube.com/watch?v=k2P_pHQDlp0&t=404s

[40] Md. Ileas Pramanik, Raymond Y.K. Lau, Md. Abul Kalam Azad, Md. Sakir Hossain, Md. Kamal Hossain Chowdhury, B.K. Karmaker, Healthcare informatics and analytics in big data, Expert Systems with Applications, Volume 152, 2020, 113388, ISSN 0957-4174, https://doi.org/10.1016/j.eswa.2020.113388.

[41] Shae, Z.; Tsai, J.J. On the design of a blockchain platform for clinical trial and precision medicine. In Proceedings of the 2017 IEEE 37th International Conference on Distributed Computing Systems (ICDCS), Atlanta, GA, USA, 5–8 June 2017.

[42] Alhadhrami, Z.; Alghfeli, S.; Alghfeli, M.; Abedlla, J.A.; Shuaib, K. Introducing blockchains for healthcare. In Proceedings of the 2017 IEEE International Conference on Electrical and Computing Technologies and Applications (ICECTA), Ras Al Khaimah, UAE, 21–23 November 2017.

[43] Fernández-Caramés, T.M.; Fraga-Lamas, P. A Review on the Use of Blockchain for the Internet of Things. IEEE Access 2018, 6, 32979–33001.

[44] Kuo, T.-T.; Hsu, C.-N.; Ohno-Machado, L. ModelChain: Decentralized Privacy-Preserving Healthcare Predictive Modeling Framework on Private Blockchain Networks. arXiv 2016, arXiv:1802.01746 .

[45] Zheng, Z.; Xie, S.; Dai, H.; Chen, X.; Wang, H. An overview of blockchain technology: Architecture, consensus, and future trends. In Proceedings of the 2017 IEEE International Congress on Big Data (BigData Congress), Honolulu, HI, USA, 25–30 June 2017.

[46] Investopedia "Blockchains".

[47] Hertig, A. Blockchain's Once-Feared 51 Percent Attack Is Now Becoming Regular. June 2018.

[48] Linn, L.A., Koo, M.B. Blockchain for health data and its potential use in health it and health care related research. In ONC/NIST Use of Blockchain for Healthcare and Research Workshop; ONC/NIST: Gaithersburg, MD, USA, 2016.

[49] Esposito, C.; De Santis, A.; Tortora, G.; Chang, H.; Choo, K.K. Blockchain: A Panacea for Healthcare Cloud-Based Data Security and Privacy? IEEE Cloud Comput. 2018, 5, 31–37. [CrossRef]

[50] Bennett, B. Blockchain HIE Overview: A Framework for Healthcare Interoperability. Telehealth Med. Today 2017, 2.

[51] Pirtle, C.; Ehrenfeld, J. Blockchain for Healthcare: The Next Generation of Medical Records? J. Med. Syst. 2018, 42, 172.

[52] McKinlay, J. Blockchain: Background Challenges and Legal Issues; DLA Piper Publications: London, UK, 2016.

[53] Boulos, M.N.; Wilson, J.T.; Clauson, K.A. Geospatial blockchain: Promises, challenges, and scenarios in health and healthcare. Int. J. Health Geogr. 2018, 25.

9

Blockchain in Healthcare: A Systematic Review and Future Perspectives

Arpneek Kaur[1], Sandhya Bansal[2], and Vishal Dattana[3]

[1]Department of Computer Science and Engineering, Maharishi
Markandeshwar Engineering College, Maharishi Markandeshwar
Deemed to be University, India;
Email: arpneek.kaur@mmumullana.org
[2]Department of Computer Science and Engineering, Maharishi
Markandeshwar Engineering College, Maharishi Markandeshwar
Deemed to be University, India;
Email: sandhyabansal@mmumullana.org
[3]Department of Computer Engineering, Middle East College,
Sultanate of Oman;
Email: vishaldattana@gmail.com

Abstract

Healthcare, being a basic necessity and a fundamental human right, is a
paramount discipline of research today. During the past decade, blockchain
technology has gained an increased interest from academic scholars for its
use in the healthcare sector. Applications such as healthcare supply chain,
identification of fake drugs, managing and securing electronic health
records, managing patient appointments with healthcare providers, regular
patient monitoring using various tools, and telemedicine come into picture
with respect to healthcare. A thorough survey of related research works in
this domain is required in order to gain insights into the future ideas and
recommendations for the use of blockchain technology in the healthcare
sector. In this book chapter, we aim to provide an exhaustive systematic
review of research publications related to smart healthcare applications by
the use of blockchain technology. With a thorough review and study of the
research works selected by systematic approach, this book chapter aims

to answer a set of research questions focusing on blockchain platforms, consensus methods, smart contracts, system evaluation methods, and their drawbacks. A tabular comparison of all the research articles on the basis of the research questions is provided. In the end, we give our observations of the challenges and limitations of the research works under this study, and their possible solutions. Thus, we reach up to a future scope of blockchain technology in healthcare applications.

9.1 Introduction

Healthcare is the focused area of research today as the world has witnessed widespread illnesses like the COVID-19 pandemic in the last two years. Researchers today are proposing and publishing their healthcare systems in large numbers in order to match the need of the hour. Modern healthcare systems are making use of state-of-the-art technologies in order to make diagnosis and treatment better in all ways. When it comes to modern healthcare machines, testing machines, and software, each has a common requirement that they need to store some or other sort of data. As the size of this data grows, it becomes equally important to maintain the integrity of this data, the accuracy, and access rights of this data because it is the integrity of this data which builds a trust between the participating entities – like the patients, the healthcare providers, the hospitals, the law bodies, etc. Blockchain technology comes with its own unique features like trust, integrity, immutability, non-repudiation, and authentication, which are required in all cases mentioned above. Blockchain technology became first known by its use in the Bitcoin cryptocurrency. Bitcoin is only a cryptocurrency and allows only the transactions of bitcoins. Gradually, the same technology became applicable to other real-world systems like in real estate, voting systems, insurance systems, secure chat rooms, supply chain systems, the very famous NFTs, and healthcare systems [49] not being an exception. This rise in the usage of blockchain systems has only taken a decade, owing to its very benefits which can be applicable in each and every domain. There is a huge amount of literature in the past five years containing new blockchain-based healthcare system models. Recognizing a need for an exhaustive review of all such quality works aims to provide the same in this book chapter. The field of blockchain technology is very diverse with its varied choices of platforms, consensus rules, and system features each providing different sets of characteristics. This book chapter contributes to the academicians by providing a comparative picture of the use of different elements of blockchain technology in recent literature works related to healthcare. At the end, it also derives some open challenges and future scope in this field. Thus, it shows a path to new blockchain-enabled research works in healthcare domain. The remaining part of this article has been organized as follows. Section 9.2 provides some applications and challenges in the modern healthcare industry which can be successfully dealt with the blockchain technology. In Section 9.3, an introduction to the blockchain technology and its important elements has been provided. Next, we provide the research methodology for selection of literature works in Section 9.4. Section 9.5 provides a theoretical literature review of the selected articles. In Section 9.6, we mention the findings of this exhaustive review based on the research questions in the form of tables and pie charts. Section 9.7 provides some open challenges and future scope for the use of blockchain technology in healthcare, followed by conclusion in Section 9.8.

9.2 Healthcare Challenges vs. Blockchain Opportunities

Today, blockchain technology is being successfully employed in numerous practical applications of varied domains, healthcare being one of them. Within the healthcare domain, there can be seen various applications of blockchain technology as shown in Figure 9.1.

Modern healthcare applications pose a lot of challenges due to the very evolving nature of technology today. Each new technique comes with its own set of challenges when it comes to practical implementation. Blockchain technology has evolved as a rescue to some of the important challenges in healthcare domain. In Figure 9.2, we mention some of the important healthcare challenges and explain how blockchain helps in solving them.

9.2.1 Interoperability

Patients as of now have no control or own responsibility for their clinical information, which makes it almost outlandish for data to be divided between various medical clinics. Moreover, security, trust, credibility, and responsibility issues demolish the circumstance for patients. Blockchain gives an option in contrast to the current detached cycle for gathering and putting away persistent information. A stage that depends on blockchain and is divided among patients, emergency clinics, and other medical care parties contains permanent, secure, and recognizable information, tackling the issue with legitimacy. Simultaneously, the information put away on a blockchain

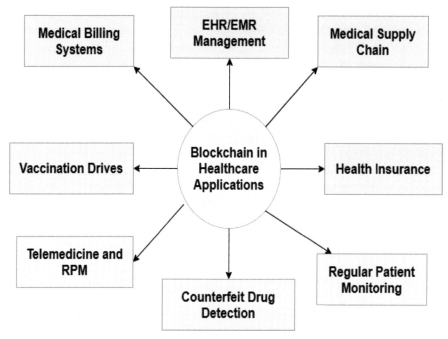

Figure 9.1   Blockchain in healthcare applications.

can undoubtedly be pseudonymized, tackling the issue with patient ID. In the blockchain-based model, patients and their clinical information are at the focal point of medical care. This implies that more control and proprietorship is allowed to patients – they get to conclude who can get and peruse their clinical information containing ailment, medicine, lab tests, medical history, etc., in the form of EHRs [17]. Simultaneously, medical clinics can all the more effectively access shared data and give better consideration and worth to the patient. With the ascent of IoT and information accumulated from wearables or portable applications, the issue with information interoperability turns out to be increasingly significant [18]. Blockchain is a chance to settle it.

9.2.2 Tampering/data security

Since there is no central entity in a blockchain system, an agreement among all members should be reached before any information gets refreshed or added to the framework. Saving every one of the details behind the agreement calculations, this implies that the framework stores changeless path of all collaborations at any point made. One of the serious issues in medical care on a worldwide scale is the absence of all around perceived patient identifiers

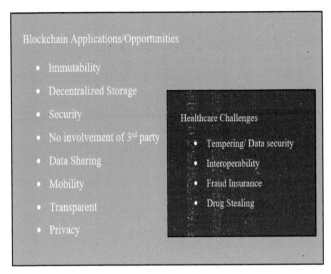

Figure 9.2   Healthcare challenges vs. blockchain opportunities.

[53]. Indeed, even arrangements like EHR cannot totally take care of this issue and have, previously, brought about jumbling of patient records. Blockchain, then again, gives the essential components to producing secure and reliable clinical records, which are difficult to get undermined or altered on the grounds that every communication in the framework at any point made passes on a review trail so that everyone might see [16]. Essentially, blockchain could tackle one more large issue in medical care, for example, oversee clinical preliminary information, while consenting to guidelines. Since clinical preliminaries as a rule require long responsibility and weighty venture, there is a danger of having the eventual outcomes changed or altered for different uses like fraudulent insurance claims [83], misuse of drugs [74], etc. In any case, defiling preliminary information is almost inconceivable in a universally open information stage that gives changeless path, all things considered. Besides, preliminary members can be compensated for following the arrangement intently, guaranteeing a drawn out responsibility.

9.2.3 Insurance fraud

Debasing clinical preliminary outcomes is not the main misrepresentation from which the medical care industry endures. An issue on a more noteworthy scale is battling protection extortion [85]. Blockchain can assist with forestalling bogus case sections by patients or potential suppliers by basically putting away all exchanges safely, giving detectability of every exchange.

9.2.4 Drug stealing

Fake medications represent a gigantic danger to medical care. Not exclusively does taking and faking drugs cause gigantic monetary misfortunes to makers, yet more critically, it might change the first arrangement and elements of an item, straightforwardly (and at times lethally) influencing the patient's wellbeing [74] [76]. Blockchain can possibly take care of this issue by giving a straightforward, shared dataset, where makers and drug providers can follow and confirm drug beginning. Indeed, the Drug Supply Chain Security Act (DSCSA), due to become successful in 2024, as of now proposes that advanced following and recognizability of medications will be needed sooner rather than later.

9.3 Introduction to Blockchain

Blockchain stands for a distributed ledger shared on a fully decentralized peer-to-peer network of nodes. Each node of the network stores its own copy of this distributed ledger [15]. Any changes to this ledger are verified and validated by each of the nodes in the network. In case a node becomes inactive for a period, it can update its copy of the ledger from any of the peer nodes. The consistency of this shared ledger is maintained with the help of a distributed consensus mechanism. Each active node validates a transaction before it can be written to the ledger. Once a part of the ledger, it becomes immutable and remains in the blockchain logs. The data in the blockchain is written in the form of blocks. The blocks are linked together to form a chain by the use of previous hash in each block. With the help of this structure, the data in the blockchain can never be tampered with. In order to change a single transaction in any block of the blockchain, all the further blocks need to be changed, which is a difficult task. The longer the chain of blocks grows, the more secure it becomes.

9.3.1 Structure of blocks and blockchain

The blocks in the blockchain contain the transaction information which has been held over the data. Each block contains various fields with different purposes. The block format may also vary from platform to platform and from application to application. There are some common important fields which are there in the blocks. The structure of a block and the blockchain has been depicted in Figure 9.3. The transactions field contains the data related

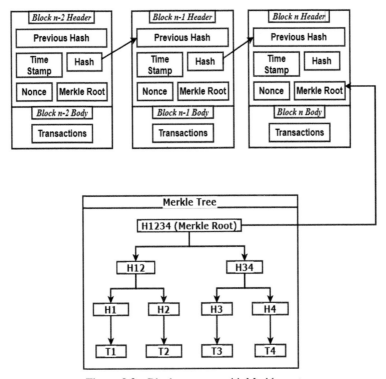

Figure 9.3   Block structure with Merkle root.

to the transactions held on the network between different parties or on the data itself. Each block may contain "*n*" number of transactions selected with the help of consensus protocols [69]. The timestamp field contains the time when the block was created. A block created and written to the blockchain once cannot be changed again. The value of nonce is calculated by extensive computations in order to ensure a particular difficulty level of hash. The hash and previous has fields in the block to ensure the immutability of the blocks. Any minute change in the block data immediately reflects in the hash value. If a malicious user tries to add his block inside the blockchain, he needs to calculate the nonce values and hashes for all the subsequent blocks with at least 51% of the nodes of the network. The transactions contained inside a block are arranged in the form of a Merkle tree. Merkle tree contains the hash of each transaction at its leaf nodes. The nodes at each higher level contain the combined hash of previous hashes [61]. Merkle tree facilitates in easier verification and validation of individual transactions without considering the other transactions in the block.

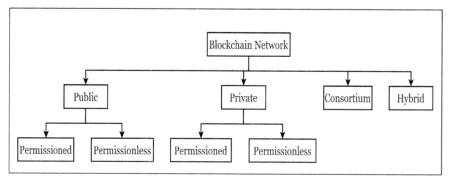

Figure 9.4   Types of blockchain networks.

The root of the Merkle tree, called the Merkle root, is stored within the header of the block. This maintains the integrity and consistency of each and every transaction in the blockchain.

9.3.2 Types of blockchain networks

Blockchain networks can be broadly classified into public and private types according to the accessibility and permissions to join the blockchain network. Other types of networks include permissioned, consortium, hybrid which can be viewed like a hybrid between the public and private networks, each with their own variations as shown in Figure 9.4. Each type of blockchain network has its own set of characteristics, owing to various advantages and disadvantages. One must choose a type of network according to the need of the application for which blockchain is to be employed.

A public blockchain is free to join for anyone. There are full nodes, miner nodes, and wallet nodes in a network according to their computation powers and capacity. There are no constraints on the roles of any node [79]. Some business firms do not like to expose their sensitive data to the world wide nodes on the public blockchain network. In a private network, a business firm decides who will join the network, what type of transactions will be allowed, and what consensus protocols will be used [19]. Permissioned blockchain networks can be employed with a special access control layer which provides particular roles and access rights to the users in the network [2]. Permissioned blockchain networks can be viewed as a mid-way between fully private and fully public blockchains. A consortium blockchain is typically a blockchain network formed by multiple organizations related to the same type of business as in [91] and [93]. There are hybrid blockchain networks that use both the public and private blockchains for different types of data as in [70] and [90].

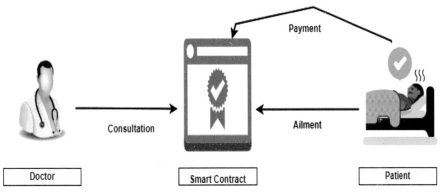

Figure 9.5 Smart contracts.

9.3.3 Consensus algorithms

The blockchain network consists of a large number of nodes spread across different locations. Each node has its own copy of the distributed ledger. Whenever there is an update for the ledger, it has to be updated the same with all of the nodes on the network. That is why there is a need for some mechanism by which all the nodes of the network can come to an agreement regarding the new block to be added to the blockchain. This mechanism is called "consensus". A consensus mechanism must not be centralized which will go against the integral properties of blockchain [35]. The nodes of the blockchain network must have equal probability of being able to add their blocks in their blockchain. There are different algorithms used in the blockchains for distributed consensus.

9.3.4 Smart contracts

Smart contracts are small programs that can reside on the blockchain network. The execution of this program code is automatically triggered when certain predefined conditions are met. The smart contracts execute on the nodes of the blockchain network rather than on a central server. Thus, once migrated and deployed to the blockchain network, smart contracts themselves become immutable, and, thus, no party can change and validate the contractual conditions without the consent of the other parties involved [46]. Thus, smart contracts act as a smart substitute for traditional paper contracts and also eliminate the need of a third party such as a lawyer. Smart contracts can be written in any programming language. Some of the common programming languages used for smart contracts are Solidity, Go, Java, JavaScript, Python, etc. Smart contracts provide innumerable benefits like speed, autonomy, accuracy, transparency, and cost savings. Figure 9.5 shows an example of a healthcare smart contract between a doctor and a patient.

9.4 Research Methodology

As it is a systematic review, it is necessary to select the most relevant papers related to blockchain in healthcare in the year 2017–2021. Various search strategies were formed for this task. This review article has been focused on the following review questions:

- **RQ1:** What type of blockchain network and blockchain platform has been used by researchers in their healthcare applications?
- **RQ2:** What are the important consensus algorithms used in blockchain-based healthcare systems?
- **RQ3:** What proportion of research works have built smart contracts in their healthcare systems and which programming languages have they used?
- **RQ4:** What are the evaluation methods used by researchers for analyzing their models?
- **RQ5:** What are the drawbacks of the existing blockchain systems related to healthcare and what are the future perspectives?

The article selection procedure is shown in Figure 9.6. First, the five most relevant databases, IEEE-Xplore, Science Direct, Springer, Google Scholar, and PubMed, were searched using the various keywords like "Blockchain AND EHR", "Blockchain AND EMR", "Blockchain AND Medical Supply Chain, "Blockchain AND Health Insurance", "Blockchain AND Drug Supply Chain", etc. A total of 1980 articles were fetched. After this, a filter on the basis of title scan, removal of duplicates, magazine, conference proceedings, conference papers with pages less than 5, book chapters, and exclusion of review papers were applied and a total 127 papers were selected. Next, this count is reduced to 105 by applying a filtering process on the basis of citation count and quality check. Finally, selection criteria based on abstract scan and review questions were conducted that led to selection of final 90 articles for review.

9.5 Literature Review

The authors in [1] had designed a smart healthcare system with the integration of blockchain 3.0 and healthcare 4.0. The designed network has been implemented in Ethereum network along with other programming languages

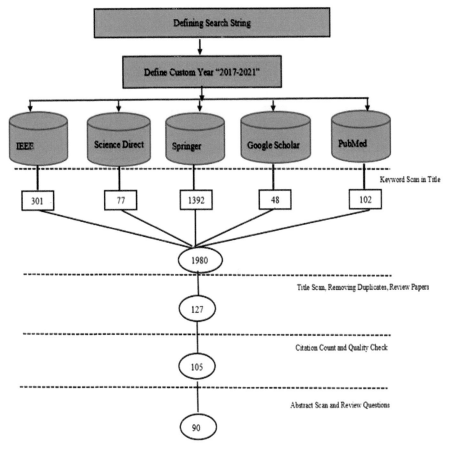

Figure 9.6 Article selection procedure.

and tools like solidity web3.js. Further, to handle the challenge of trust building, smart contracts have been used. The work [2] presents a blockchain-based secure, efficient, and easily maintainable framework for medical data access between patient and doctor. The proposed framework allows the patient to have a full control of its data. The work [13] also develops a framework for secure, robust, stable, and patient-centric health information exchange. Distributed ledgers and smart contracts are used for the purpose. The work [43] develops a patient-centric application named PACEX that allows the sharing of EHR data. Ethereum blockchain and smart contracts have been used. The authors in [3] extended work of [9] and designed a hybrid architecture using blockchain and edge node for EHR data management. This architecture authenticates the user signature by using attribute-based multi-signature

(ABMS) scheme and encrypts EHR data by using multi-authority attribute-based encryption (ABE) scheme. Hyperledger Fabric platform and Hyperledger Ursa library have been used for implementation of the scheme respectively. The authors in [4] first proposed a cross blockchain based EHR storage solution to handle the interoperability issue of blockchain in healthcare. A relay-chain as a service (Raas) has been used for this purpose and it allows the deletion of the EHR information by the patient when it enters the different hospitals. The work [5] develops a blockchain-based platform for enhanced EHR in the area of diabetes as it is the hastily growing disease in today's world. This platform generates an alert message to patients and caretakers in case of emergency. The work [6] proposes a blockchain- and hash-based new strategy for enchaining the security in distributed EHR. This strategy uses collision-resistant hash function for fast generation of unique patient IDs, PoW for reducing scalability, smart contracts, and cryptography techniques for consent management. The works [7] and [40] issue the scalability of blockchain in EHR. For increasing the scalability, the proposed architecture uses multi-hyperledgers: one for patients and another for health institutes. To store a large amount of data and more request fulfillment, each channel maintains its own ledger. The work [8] integrates AI with blockchain for the more secure EHR, efficient data management, and reliable diagnosis of disease. A constraint goal model (CGM) is used to stimulate the system requirement. The work [50] issues the scalability of NHIA's NMR used in Taiwan. The developed framework can integrate the NHIA's NMR data with EMR and EHR. The work [10] develops a consent model for data sharing and access control. Smart contracts are used for representing consent. This model uses data use ontology (DUO) and automatable discovery and access matrix (ADA-M) for modeling consent of users and description of queries from requesters, respectively. The Ethereum platform is used for data validation. The papers [11] and [55] propose a scheme for secure multi-authority EHR in cloud storage environments. This scheme uses an attribute signcryption scheme for achieving security. Moreover, the use of verifiable outsourcing computational mechanisms reduces the computational burden of the user. In addition to this, [22] develops an efficient authentication scheme for resolution of centralized problems of the cloud. The work [12] uses fuzzy decision models for evaluation of blockchain models for secure and trustworthy EHR. This evaluation helps in the selection of the most suitable blockchain for trustworthy EHR. The work [14] uses zk-proofs to authenticate the beneficiaries. However, [29] uses keyless signature infrastructure for securing EHR data over the cloud. The proposed framework is evaluated and compared with existing frameworks on different parameters. The work [42] uses a revocable attribute scheme to protect privacy. Attribute signing key is calculated with the combination of attribute master key and update master key. The paper [16] proposes a multi-tier architecture for privacy preservation. This novel architecture uses public blockchain between different EHR cloud providers and other between patient sensors and patient systems. The papers [28] and [33] use cryptographic functions for maintaining the pseudonymity of the patient data. The paper [49] uses various utility functions to evaluate privacy preserving frameworks. Multiple smart contracts are used for evaluation of access control and transaction evaluation. The paper [35] uses blockchain for securing decentralized EHR sharing. The work [18] proposes a hybrid architecture that combines blockchain and edge nodes to facilitate EHR access control. This architecture maintains a tamper proof log of the access events. Hyperledger Fabric is used for the evaluation of the hybrid architecture. In addition to this, [19] and [44] also propose a novel EHR framework for trustworthy access control mechanisms between various users of EHR. Ethereum is used for prototype implementation. Furthermore, [20] focuses on patient controlled sharing schemes instead of third parties. Some medical institutes are used for storing the PESchain. These are used for storing the hash values of the encrypted patient record. Stealth authorization scheme is constructed for privacy preservation and sharing of data. The work [21] addresses the problem of privacy preservation and security over cloud sharing of EHR records. Searchable encryption and conditional proxy re-encryption based scheme is developed for the resolution of problems. Proof of authorization is used as a consensus protocol. The paper [24] presents a novel audit mechanism for exchanging e-health data across the borders in Europe, while [47] proposes an architecture for EHR data in Peruvian health organizations. For off-chain storage of medical data and providing logs of transactions, laws maintained by Peruvian are followed. The paper [48] proposes a framework for sharing and management of cancer patients. The data validation is done on the data of Stony Brook University Hospital. The paper [25] presents a blockchain-based solution for secure EHR inter-exchange. Security analysis is done in terms of inside and outside threats on the system. The work [41] proposes a permissioned blockchain for cross-organizational data sharing. The paper [26] proposes a novel and secure scheme named MedSBA, an attribute-based encryption scheme for sharing of EHR. Private blockchain is utilized. The work [27] develops real, operational, and secured PHRs. Three design principles were developed for secure PHR. The work [36] proposes a certificate-less scheme and elliptical curve cryptography scheme for the security of data. The paper [30] integrates Hyperledger blockchain with deep learning for secure data management and diagnosis model, respectively. The developed model works in five different stages. SIMON block cipher technique is applied for encryption of data and variational autoencoder is applied for detection of disease. The work [31] utilizes smart contracts for sharing of data between patients, doctors, and caretakers. Modified Merkle tree data structure is used for secure and fast access of data. Cryptographic hash functions are used for security of data. The paper [32] uses attribute and identity-based encryption for security of data over the cloud. The paper [37] develops a novel model MedRec for data access and permission management of medical data. The work [39] proposes a granular access control mechanism that can give different levels of authorization while maintaining compatibility with underlying blockchain architecture. The work [45] focuses on the EHR data of emergency wards of the hospitals. This work uses the P2P network for analyzing data transmission of smart contracts. The paper [46] builds a unified solution for migrating the existing EHR. Off-chain storage of assets is controlled through patient and policy transactions.

9.6 Discussion

We have already provided a brief literature review of the selected articles. We can successfully derive the answers to our research questions based on the survey. In this section, we head towards a discussion of our findings on the basis of research questions one by one.

9.6.1 RQ1: What type of blockchain network and blockchain platform has been used by researchers in their healthcare applications?

Table 9.1 Types of blockchain networks in recent works

Type of network	References
Public	[16], [51], [54], [62], [63], [68], [69], [72], [76], [79], [80], [81], [83], [86], [87]
Private	[9], [19], [23], [24], [25], [27], [30], [34], [37], [39], [43], [46], [50], [52], [59], [73], [98]
Hybrid	[3], [18], [26], [57], [66], [67], [70], [75], [90]
Consortium	[20], [21], [22], [32], [38], [40], [44], [47], [56], [58], [61], [77], [82], [85], [88], [91], [93], [98]
Permissioned	[2], [10], [28], [31], [33], [34], [35], [36], [42], [44], [46], [48], [55], [56], [60], [64], [65], [71], [78], [84], [89], [92], [95], [96], [97]

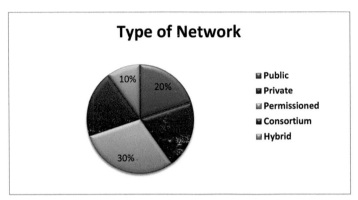

Figure 9.7 Comparison of network types in recent works.

Table 9.2 Blockchain platforms in recent works

Blockchain platform used	References
Hyperledger (Fabric/Burrow/Composer/Iroha/Sawtooth)	[3], [18], [27], [30], [34], [35], [44], [46], [47], [48], [56], [64], [65], [67], [78], [84], [91], [92], [94], [95], [96], [97]
Ethereum	[2], [10], [19], [20], [21], [25], [33], [35], [36], [37], [38], [40], [43], [50], [51], [54], [55], [57], [60], [68], [69], [72], [76], [79], [80], [81], [83], [87], [88], [91], [94]
Developed by authors	[70], [71], [73], [85], [89]
Wanchain	[23]
OPNET	[26]
Multichain	[24]
IOTA	[40]
Cardano	[40]
Bitcoin	[40], [63]
NEM	[75]
QuarkChain	[90]
Gcoin	[93]

The type of platform is chosen by the researchers according to the requirements of the system. Table 9.1 categorizes the reviewed literature works based on the type of network, viz. public, private, hybrid, consortium, and permissioned. Figure 9.7 depicts that researchers prefer to use permissioned systems above all because they provide benefits of access control and role

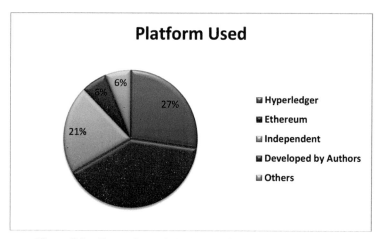

Figure 9.7　Comparison of blockchain platforms in recent works.

definition. However, we can see that there are lesser number of research works with hybrid systems, which can be viewed as a future scope in this direction owing to the very benefits of public–private hybrids. In Table 9.2 as well as Figure 9.7, we can very well notice that Ethereum is the most widely used platform by the researchers in healthcare domain, with Hyperledger being the second one. However, more research works are needed to find out the performance of varied platforms such as IOTA, Cardano, GCoin, etc. It has also been noticed that only a 7% of researchers have developed their own platforms.

9.6.2 RQ2: What are the important consensus algorithms used in blockchain-based healthcare systems?

Consensus methods are a very important element of a blockchain system. The selection of a good distributed consensus method leads to cost and time savings in the network. The consensus methods used by selected research works have been summarized in Table 9.3. Also, Figure 9.8 shows that proof of work is the most used consensus algorithm till date. More research works are needed with other consensus methods that have lesser cost and computation requirements. Also, there is more scope of new consensus algorithms to be built by authors in future works.

Table 9.3   Consensus algorithms used in recent works

Consensus algorithm	References
Proof of work	[2], [10], [33], [37], [51], [54], [55], [62], [68], [69], [72], [76], [79], [80], [81], [91], [93]
Proof of authority	[21], [40], [50], [60], [90]
Delegated proof of stake	[38], [52], [70], [94]
Created by authors	[30], [32], [58], [85], [95], [98]
Practical byzantine fault tolerance	[26], [44], [48], [61], [64], [84], [91], [94], [97]
Proof of concept	[19], [47]
Proof of stake	[86]
Proof of reputation	[63]
Proof of importance	[75]
Byzantine fault tolerance	[83], [89], [92]
Istanbul byzantine fault tolerance	[35]
Kafka	[65]
Galaxy consensus	[23]

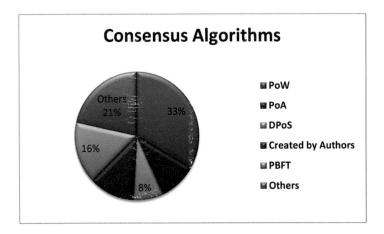

Figure 9.8   Comparison of consensus algorithms in recent works.

9.6.3 RQ3: What proportion of research works have built smart contracts in their healthcare systems and which programming languages have they used?

Smart contracts, being an important element of the newer versions of blockchain, are employed in various healthcare systems for maintaining the integrity, access control, and other security features of the system. However, in Figure 9.9, we see that a huge proportion of about 42% research works are not proposing any smart contracts in their models, which is an important requirement for future systems.

Table 9.4   Smart contract languages in recent works

Programming language	References
Solidity	[2], [10], [19], [21], [25], [33], [35], [37], [38], [43], [50], [51], [55], [57], [60], [68], [69], [72], [76], [79], [80], [81], [86], [95]
ChainCode	[46], [47], [48], [64], [65], [67], [84], [92], [96], [97]
Java Script	[18], [27], [78]
Python	[31], [83], [88]
Swift	[75]
Rust	[3]
Go	[30]

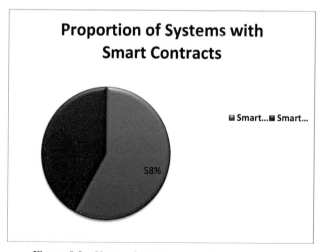

Figure 9.9   Usage of smart contracts in recent works.

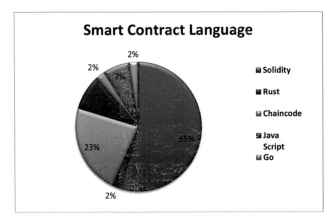

Figure 9.10 Comparison of smart contract languages used in recent works.

Table 9.4 summarizes the languages selected by the remaining of the systems for their smart contracts. In Figure 9.10, we can notice that Solidity, being the only smart contract language for Ethereum, is the most widely used, whereas there is a huge future scope of newer smart contract languages like Rust, Go, etc.

9.6.4 RQ4: What are the evaluation methods used by researchers for analyzing their models?

There are different evaluation methods used in research works for analysis of the proposed models. Our finding as shown in Table 9.5 and Figure 9.11 is that a huge proportion, i.e., 46% of works are using performance metrics such as latency, throughput, communication cost, computation cost, transaction time, data size, etc., for the evaluation of their proposed systems. However, lesser researchers are interested towards survey-based evaluation methods, which should also be opted because blockchain, being a new technology, needs to be testified by the end users of the system in order to gain popularity.

9.6.5 RQ5: What are the drawbacks of the existing blockchain systems related to healthcare and what are the future perspectives?

We have come across some common drawbacks of existing systems which need to be addressed in the future works. Table 9.6 summarizes some of the drawbacks identified in the existing systems. We can see that data privacy issues and scalability issues are the most common among the articles under

Table 9.5   System evaluation methods used in recent works

Evaluation method	References
System analysis by discussion of different features like privacy, access control, immutability, transparency, interoperability, etc., depending on proposed models	[2], [20], [23], [25], [28], [37], [39], [42], [44], [46], [47], [48], [52], [54], [55], [58], [61], [66], [67], [68], [69], [70], [72], [75], [76], [77], [79], [80], [82], [84], [86], [89], [92], [93]
Simulation on Hyperledger Composer/Fabric/Calliper/ Ganache/OPENT/Bitcoin/ Rinkeyby/AWS/Others	[10], [18], [19], [20], [21], [25], [26], [34], [36], [39], [55], [56], [57], [63], [70], [81], [83], [90], [94], [95], [97]
Performance metrics (latency/ throughput/communication cost/ computation cost/transaction time/ data size/gas spent, etc., depending on proposed models) evaluation by experimentation on various tools	[3], [10], [16], [18], [19], [20], [21], [26], [27], [29], [30], [31], [33], [35], [36], [39], [40], [42], [43], [44], [46], [47], [51], [52], [54], [55], [56], [58], [59], [60], [61], [62], [63], [65], [67], [69], [70], [71], [76], [78], [79], [80], [81], [85], [86], [88], [90], [91], [95], [96], [98]
Theorem-based proof of system security features	[22], [50], [59], [90]
Survey-based system evaluation	[27], [60], [97]

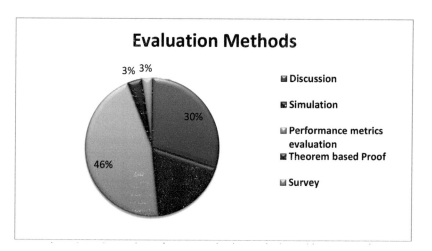

Figure 9.11   Comparison of system evaluation methods used in recent works.

Table 9.6   Common drawbacks found in recent works

Common drawbacks identified	References
Data privacy issue	[9], [10], [18], [22], [27], [40], [47], [51], [60], [66], [68], [72], [76], [78], [79], [87], [88], [89], [92], [93], [95]
Scalability problem	[2], [3], [21], [26], [31], [36], [37], [42], [50], [52], [55], [57], [62], [65], [68], [69], [72], [76], [79], [80], [83], [86], [87], [88], [92]
No standard performance metrics have been evaluated	[2], [23], [24], [28], [37], [38], [48], [57], [68], [75], [82], [83], [84], [89], [92], [93], [94], [97]
Storage and access control is not fully decentralized	[20], [24], [27], [43], [46], [61], [65], [75], [96]
High costs of computation	[2], [33], [60], [81], [92]
No user authentication method	[19], [29], [30], [39]
Decentralized consensus has not been built	[39], [42], [43], [46], [59], [67], [71], [73]
No smart contracts built	[9], [16], [20], [22], [23], [24], [26], [28], [29], [32], [36], [39], [42], [44], [52], [54], [58], [59], [61], [62], [66], [70], [71], [73], [77], [82], [85], [87], [89], [90], [93], [98]

review. In the next section, we have mentioned a detailed set of challenges and future scope which have been identified during this survey.

9.7 Challenges and Future Scope

Besides the potential benefits of blockchain technology to the healthcare industry, significant challenges still exist. We have already discussed the common drawbacks in the previous research works. Some important challenges identified during this review have been summarized below along with some existing systems which have addressed those challenges.

9.7.1 Lack of standardization

The implementation of blockchain-based healthcare systems still lacks in standardization as existing solutions are using different storage methods, consensus algorithms, encryption techniques, and different networks. Consequently, the comparison between these implementations becomes difficult. So there is a need for the evolution of some standards that can help in comparing different solutions. Some systems like [30] and [71] have used standard existing datasets for evaluating and comparing their systems in order to address this problem.

9.7.2 Scalability

As the number of blocks in the application increases, the system becomes computationally slower, and it requires more computational resources and this situation may be worse in an IoT environment where the sensor devices are equipped with limited computational power just enough for performing transmission and receiving data. However, some recent systems [18], [35], [51], have successfully addressed the scalability problem in their models.

9.7.3 Latency

Latency plays an important role in the healthcare sector wherein the personal satisfaction of patients profoundly relies upon continuous medical care frameworks, particularly when healthcare data is gathered through the IoT gadgets and should be recorded into blockchain frameworks. The issue becomes critical when the EHRs are re-evaluated on the brought together distributed storage frameworks. This is because the cloud services may not be accessible at the area of the medical services suppliers, bringing about impressive network latency to re-evaluate the data. It is worth taking note of that network latency has a perceptible effect on fostering a versatile blockchain-based framework. On the other hand, some systems like [31] and [35] have made a successful attempt to address this problem.

9.7.4 Privacy of outsourced data

Data storage is a crucial issue in the blockchain technology that can be handled either on-chain or off-chain. Possible solution followed by existing work is to outsource the healthcare data on the cloud and management of this into the cloud. However, storing and sharing data on cloud computing is neither secure nor trustable. Moreover, the cloud service providers may be curious about the medical information and overstep the patients' privacy by revealing the healthcare records and other private information. This curious nature of the service providers put the data at risk. However, systems like [28] and [52] have addressed the privacy problem in their models.

Conclusion

In this book chapter, we have provided an exhaustive review of 90 quality articles selected by the SR approach. This review has led to some important findings based on our review questions. There is an enormous future scope of new platforms, consensus methods, and programming languages in the healthcare systems enabled by blockchain technology. We see that a huge proportion of the research works are focused on a limited type of platforms. We can also see that the existing systems have some common drawbacks which need to be addressed in the future works. The findings of this book chapter contribute towards a path for new research works in this domain. We can conclude here by saying that a huge future scope of the blockchain technology can be seen in healthcare applications.

References

[1] A. Kumar, R. Krishnamurthi, A. Nayyar, K. Sharma, V. Grover, E. Hossain, 'A novel smart healthcare design, simulation, and implementation using healthcare 4.0 processes', IEEE Access, 8, 118433-118471,2020

[2] J. Vora, A. Nayyar, S. Tanwar, S. Tyagi, N. Kumar, M.S Obaidat, J.J Rodrigues, 'BHEEM: A blockchain-based framework for securing electronic health records', In 2018 IEEE Globecom Workshops (GC Wkshps), pp. 1-6, IEEE, Dec., 2018.

[3] H. Guo, W. Li, E. Meamari, C. Shen, M. Nejad. 'Attribute-based multi-signature and encryption for ehr management: A blockchain-based solution', In 2020 IEEE International Conference on Blockchain and Cryptocurrency (ICBC), pp. 1-5, IEEE, May, 2020.

[4] S. Cao, J. Wang, X. Du, X. Zhang, X. Qin, 'CEPS: A Cross-Blockchain based Electronic Health Records Privacy-Preserving Scheme', In ICC 2020-2020 IEEE International Conference on Communications (ICC), pp. 1-6, IEEE, June, 2020.

[5]   C. Priya, 'Enabling the efficiency of Blockchain Technology in Tele-Healthcare with Enhanced EMR', In 2020 International Conference on Computer Science, Engineering and Applications (ICCSEA), pp. 1-6, IEEE, March 2020.

[6]   A.I. El Sayed, M. Abdelaziz, M.H. Megahed, M.H.A. Azeem, 'A New Supervision Strategy based on Blockchain for Electronic Health Records', In 2020 12th International Conference on Electrical Engineering (ICEENG), pp. 151-156, IEEE, July 2020.

[7]   A. Fernandes, V. Rocha, A.F da Conceição, F. Horita, 'Scalable Architecture for sharing EHR using the Hyperledger Blockchain', In 2020 IEEE International Conference on Software Architecture Companion (ICSA-C), pp. 130-138, IEEE, March 2020.

[8]   Y. Wehbe, M. Al Zaabi, D. Svetinovic, 'Blockchain AI framework for healthcare records management: Constrained goal model', In 2018 26th Telecommunications Forum (TELFOR) pp. 420-425, IEEE, Nov. 2018.

[9]   R. Guo, H. Shi, Q. Zhao, D. Zheng, 'Secure attribute-based signature scheme with multiple authorities for blockchain in electronic health records systems'. IEEE access, 6, 11676-11686, 2018.

[10] V. Jaiman, V. Urovi, 'A consent model for blockchain-based health data sharing platforms', IEEE Access, 8, 143734-143745, 2020.

[11] X. Yang, T. Li, W. A. Chen, C. Wang, 'A blockchain-assisted verifiable outsourced attribute-based signcryption scheme for EHRs sharing in the cloud', IEEE Access, 8, 170713-170731,2020.

[12] M. Zarour, M. T.J Ansari, M. Alenezi, A.K Sarkar, M. Faizan, M, A. Agrawal, R. A. Khan, 'Evaluating the impact of blockchain models for secure and trustworthy electronic healthcare records', IEEE Access, 8, 157959-157973, 2020.

[13] Y. Zhuang, L. R. Sheets, Y.W Chen, Z. Y. Shae, J.J Tsai, C.R Shyu, 'A patient-centric health information exchange framework using blockchain technology' . IEEE journal of biomedical and health informatics, 24(8), 2169-2176, 2020.

[14] B. Sharma, R. Halder, J. Singh, 'Blockchain-based interoperable healthcare using zero-knowledge proofs and proxy re-encryption', In 2020 International Conference on Communication Systems & Networks (COMSNETS) (pp. 1-6). IEEE, Jan. 2020.

[15] A.S Chouhan, M.S Qaseem, Q.M.A Basheer, M.A Mehdia, ' Blockchain based EHR system architecture and the need of blockchain in healthcare' Materials Today: Proceedings, 2021.

[16] S. Badr, I. Gomaa, E. Abd-Elrahman, 'Multi-tier blockchain framework for IoT-EHRs systems', Procedia Computer Science, 141, 159-166, 2018.

[17] M. Shuaib, S. Alam, M. S Alam, M.S Nasir, 'Compliance with HIPAA and GDPR in blockchain-based electronic health record', Materials Today: Proceedings, 2021.

[18] H. Guo, Li, W., Nejad, C.C Shen, 'Access control for electronic health records with hybrid blockchain-edge architecture', In 2019 IEEE International Conference on Blockchain (Blockchain) pp. 44-51, IEEE, July 2019.

[19] D. C Nguyen, P. N. Pathirana, M. Ding, A. Seneviratne, 'Blockchain for secure ehrs sharing of mobile cloud based e-health systems' IEEE access, 7, 66792-66806, 2019.

[20] S. Jiang, H. Wu, L. Wang, 'Patients-controlled secure and privacy-preserving EHRs sharing scheme based on consortium blockchain' In 2019 IEEE Global Communications Conference (GLOBECOM) (pp. 1-6) IEEE, December 2019.

[21] Y. Wang, A. Zhang, P. Zhang, H. Wang, 'Cloud-assisted EHR sharing with security and privacy preservation via consortium blockchain', IEEE Access, 7, 136704-136719, 2019.

[22] F. Tang, S. Ma, Y. Xiang, C. Lin, 'An efficient authentication scheme for blockchain-based electronic health records', IEEE access, 7, 41678-41689, 2019.

[23] Y. Wang, M. He, 'CPDS: A Cross-Blockchain Based Privacy-Preserving Data Sharing for Electronic Health Records', In 2021 IEEE 6th International Conference on Cloud Computing and Big Data Analytics (ICCCBDA), pp. 90-99 IEEE, April 2021.

[24] L. Castaldo, V. Cinque, 'Blockchain-based logging for the cross-border exchange of eHealth data in Europe', In International ISCIS Security Workshop, pp. 46-56, Springer, Cham, February 2018.

[25] O. Ajayi, M. Abouali, T. Saadawi, 'Blockchain Architecture for Secured Inter-healthcare Electronic Health Records Exchange', In International

Conference on Intelligent Networking and Collaborative Systems, pp. 161-172, Springer, Cham, August 2020.

[26] S. M. Pournaghi, M. Bayat, Y. Farjami, 'MedSBA: a novel and secure scheme to share medical data based on blockchain technology and attribute-based encryption', Journal of Ambient Intelligence and Humanized Computing, 1-29, 2020.

[27] P. Meier, J. H. Beinke, C. Fitte, F. Teuteberg, 'Generating design knowledge for blockchain-based access control to personal health records', Information Systems and e-Business Management, 19(1), 13-41, 2021.

[28] A. Al Omar, M. S. Rahman, A. Basu, S. Kiyomoto, 'Medibchain: A blockchain based privacy preserving platform for healthcare data', In International conference on security, privacy and anonymity in computation, communication and storage, pp. 534-543, Springer, Cham, December 2017.

[29] G. Nagasubramanian, R. K. Sakthivel, R. Patan, A. H. Gandomi, M. Sankayya, & B. Balusamy, 'Securing e-health records using keyless signature infrastructure blockchain technology in the cloud', Neural Computing and Applications, 32(3), 639-647, 2020.

[30] N. Sammeta, L. Parthiban, 'Hyperledger blockchain enabled secure medical record management with deep learning-based diagnosis model', Complex & Intelligent Systems, 1-16, 2021.

[31] U. Chelladurai, S. Pandian, 'A novel blockchain based electronic health record automation system for healthcare', Journal of Ambient Intelligence and Humanized Computing, 1-11, 2021.

[32] H. Wang, Y. Song, 'Secure cloud-based EHR system using attribute-based cryptosystem and blockchain', Journal of medical systems, 42(8), 1-9, 2018.

[33] A. Al Omar, M. Z. A. Bhuiyan, A. Basu, S. Kiyomoto, M. S. Rahman, 'Privacy-friendly platform for healthcare data in cloud based on blockchain environment', Future generation computer systems, 95, 511-521, 2019.

[34] M. Antwi, A. Adnane, F. Ahmad, R. Hussain, M. H. ur Rehman, C. A. Kerrache, 'The Case of HyperLedger Fabric as a Blockchain Solution for Healthcare Applications', Blockchain: Research and Applications, 100012, March 2021.

[35] K. Shuaib, J. Abdella, F. Sallabi, M. A. Serhani, 'Secure decentralized electronic health records sharing system based on blockchains', Journal of King Saud University-Computer and Information Sciences, 2021.

[36] T. Benil, J. J. C. N. Jasper, 'Cloud based security on outsourcing using blockchain in E-health systems', Computer Networks 178 (2020): 107344, 2020.

[37] A. Azaria, A. Ekblaw, T. Vieira, A. Lippman, 'Medrec: Using blockchain for medical data access and permission management', In 2016 2nd international conference on open and big data (OBD), pp. 25-30, IEEE, August 2016.

[38] J. Liu, X. Li, L. Ye, H. Zhang, X. Du, M. Guizani, 'BPDS: A blockchain based privacy-preserving data sharing for electronic medical records', In 2018 IEEE Global Communications Conference (GLOBECOM), pp. 1-6, IEEE, December 2018.

[39] X. Zhang, S. Poslad, 'Blockchain support for flexible queries with granular access control to electronic medical records (EMR)', In 2018 IEEE International conference on communications (ICC), pp. 1-6 IEEE, May 2018.

[40] A. Donawa, I. Orukari, C. E. Baker, 'Scaling blockchains to support electronic health records for hospital systems', In 2019 IEEE 10th Annual Ubiquitous Computing, Electronics & Mobile Communication Conference (UEMCON), pp. 0550-0556, IEEE, October 2019.

[41] Z. Xiao, Z. Li, Y. Liu, L. Feng, W. Zhang, T. Lertwuthikarn, R. S. M. Goh, 'Emrshare: A cross-organizational medical data sharing and management framework using permissioned blockchain', In 2018 IEEE 24th International Conference on Parallel and Distributed Systems (ICPADS), pp. 998-1003, IEEE, December 2018.

[42] Q. Su, R. Zhang, R. Xue, P. Li, 'Revocable attribute-based signature for blockchain-based healthcare system', IEEE Access, 8, 127884-127896, 2020.

[43] B. Toshniwal, P. Podili, R. J. Reddy,K. Kataoka, 'PACEX: PAtient-Centric EMR eXchange in Healthcare Systems using Blockchain', In 2019 IEEE 10th Annual Information Technology, Electronics and Mobile Communication Conference (IEMCON), pp. 0954-0960, IEEE, October 2019.

[44] D. C. Nguyen, P. N. Pathirana, M. Ding, A. Seneviratne, 'Blockchain and edge computing for decentralized EMRs sharing in federated healthcare', In GLOBECOM 2020-2020 IEEE Global Communications Conference, pp. 1-6, IEEE, December 2020.

[45] W. Jiamsawat, C. Choksuchat, S. Matayong, 'Blockchain-Based Electronic Medical Records Management of Hospital Emergency Ward', In 2021 International Conference on COMmunication Systems & NETworkS (COMSNETS), pp. 674-679, IEEE, January 2021.

[46] S. Biswas, K. Sharif, F. Li, I. Alam, S. Mohanty, 'DAAC: Digital Asset Access Control in a Unified Blockchain Based E-Health System', IEEE Transactions on Big Data, 2020.

[47] A. Martínez, C. Molina, D. Subauste, 'Electronic Medical Records Management in Health Organizations using a Technology Architecture based on Blockchain', In 2020 IEEE ANDESCON, pp. 1-6, IEEE, October 2020.

[48] A. Dubovitskaya, Z. Xu, S. Ryu, M. Schumacher, F. Wang, 'Secure and trustable electronic medical records sharing using blockchain', In AMIA annual symposium proceedings, Vol. 2017, p. 650, American Medical Informatics Association, 2017.

[49] G. Wu, S. Wang, Z. Ning, B. Zhu, 'Privacy-Preserved EMR Information Publishing and Sharing: A Blockchain-Enabled Smart Healthcare System', IEEE Journal of Biomedical and Health Informatics, 2021.

[50] Y. S. Lo, C. Y. Yang, H. F. Chien, S. S. Chang, C. Y. Lu, R. J. Chen, 'Blockchain-enabled iWellChain framework integration with the national medical referral system: development and usability study', Journal of medical Internet research, 21(12), e13563, 2019.

[51] A. Shahnaz, U. Qamar, A. Khalid, 'Using blockchain for electronic health records', IEEE Access, 7, 147782-147795, 2019.

[52] X. Liu, Z. Wang, C. Jin, F. Li, G. Li, 'A blockchain-based medical data sharing and protection scheme', IEEE Access, 7, 118943-118953, 2019.

[53] S. Shamshad, K. Mahmood, S. Kumari, C. M. Chen, 'Cloud-assisted secure eHealth systems for tamper-proofing EHR via blockchain', Journal of Information Security and Applications, 55, 102590, 2020.

[54] S. Cao, G. Zhang, P. Liu, X. Zhang, F. Neri, 'Cloud-assisted secure eHealth systems for tamper-proofing EHR via blockchain', Information Sciences, 485, 427-440, 2019.

[55] L. Chen, W. K. Lee, C. C. Chang, K. K. R. Choo, N. Zhang, 'Blockchain based searchable encryption for electronic health record sharing', Future generation computer systems, 95, 420-429, 2019.

[56] S. Tanwar, K. Parekh, R. Evans, 'Blockchain-based electronic healthcare record system for healthcare 4.0 applications', Journal of Information Security and Applications, 50, 102407, 2020.

[57] S. S. R. Krishnan, M. K. Manoj, T. R. Gadekallu, N. Kumar, P. K. R. Maddikunta, S. Bhattacharya, M. J. Piran, 'A Blockchain-Based Credibility Scoring Framework for Electronic Medical Records', In 2020 IEEE Globecom Workshops (GC Wkshps, pp. 1-6, IEEE, December 2020.

[58] K. Fan, S. Wang, Y. Ren, H. Li, Y. Yang, 'Medblock: Efficient and secure medical data sharing via blockchain', Journal of medical systems, 42(8), 1-11, 2018.

[59] J. Fu, N. Wang, Y. Cai, 'Privacy-preserving in healthcare blockchain systems based on lightweight message sharing. Sensors', 20 (7), 1898, 2020.

[60] O. Gutiérrez, G. Romero, L. Pérez, A. Salazar, M. Charris, P. Wightman, 'Healthy Block: Blockchain-Based IT Architecture for Electronic Medical Records Resilient to Connectivity Failures', International Journal of Environmental Research and Public Health, 17(19), 7132, 2020.

[61] H. Zhu, Y. Guo, L. Zhang, 'An improved convolution Merkle tree-based blockchain electronic medical record secure storage scheme', Journal of Information Security and Applications, 61, 102952, 2021.

[62] R. Johari, V. Kumar, K. Gupta, D. P. Vidyarthi, 'BLOSOM: BLOckchain technology for Security of Medical records', ICT Express, 2021.

[63] R. Zou, X. Lv, J. Zhao, 'SPChain: Blockchain-based medical data sharing and privacy-preserving eHealth system', Information Processing & Management, 58(4), 102604, 2021.

[64] M. Usman, U. Qamar, 'Secure electronic medical records storage and sharing using blockchain technology', Procedia Computer Science, 174, 321-327, 2020.

[65] Z. Chen, W. Xu, B. Wang, H. Yu, 'A blockchain-based preserving and sharing system for medical data privacy', Future Generation Computer Systems, 2021.

[66] G. Tripathi, M. A Ahad, S. Paiva, 'S2HS-A blockchain based approach for smart healthcare system. In Healthcare', Vol. 8, No. 1, p. 100391, Elsevier, March 2020.

[67] C. Zhang, C. Xu, K. Sharif, L. Zhu, 'Privacy-preserving contact tracing in 5G-integrated and blockchain-based medical applications', Computer Standards & Interfaces, 77, 103520, 2021.

[68] R. M. A. Latif, K. Hussain, N. Z. Jhanjhi, A. Nayyar, O. Rizwan, 'A remix IDE: smart contract-based framework for the healthcare sector by using Blockchain technology', Multimedia Tools and Applications, 1-24, 2020.

[69] M. Sultana, A. Hossain, F. Laila, K. A. Taher, M. N. Islam, 'Towards developing a secure medical image sharing system based on zero trust principles and blockchain technology', BMC Medical Informatics and Decision Making, 20(1), 1-10, 2020.

[70] T. Zhou, X. Li, H. Zhao, 'Med-PPPHIS: blockchain-based personal healthcare information system for national physique monitoring and scientific exercise guiding', Journal of medical systems, 43(9), 1-23, 2019.

[71] A. Mubarakali, 'Healthcare services monitoring in cloud using secure and robust healthcare-based BLOCKCHAIN (SRHB) approach', Mobile Networks and Applications, 25(4), 1330-1337, 2020.

[72] T. Bocek, B. B. Rodrigues, T. Strasser, B. Stiller, 'Blockchains everywhere-a use-case of blockchains in the pharma supply-chain', In 2017 IFIP/IEEE symposium on integrated network and service management (IM), pp. 772-777, IEEE, May 2017.

[73] R. Kumar, R. Tripathi, 'Traceability of counterfeit medicine supply chain through Blockchain', In 2019 11th International Conference on Communication Systems & Networks (COMSNETS), pp. 568-570, IEEE, January 2019.

[74] V. Ahmadi, S. Benjelloun, M. El Kik, T. Sharma, H. Chi, W. Zhou, 'Drug governance: IoT-based blockchain implementation in the pharmaceutical supply chain', In 2020 Sixth International Conference on Mobile and Secure Services (MobiSecServ), pp. 1-8, IEEE, February 2020.

[75] G. Subramanian, A. S. Thampy, N. V. Ugwuoke, B. Ramnani, 'Crypto Pharmacy–Digital Medicine: A Mobile Application Integrated With Hybrid Blockchain to Tackle the Issues in Pharma Supply Chain', IEEE Open Journal of the Computer Society, 2, 26-37, 2021.

[76] R. W. Ahmad, K. Salah, R. Jayaraman, I. Yaqoob, M. Omar, S. Ellahham, 'Blockchain-Based Forward Supply Chain and Waste Management for COVID-19 Medical Equipment and Supplies', IEEE Access, 9, 44905-44927, 2021.

[77] Y. Yue, X. Fu, 'Research on Medical Equipment Supply Chain Management Method Based on Blockchain Technology', In 2020 International Conference on Service Science (ICSS), pp. 143-148, IEEE, August 2020.

[78] H. Kumiawan, J. Kim, H. Ju, 'Utilization of the Blockchain Network in The Public Community Health Center Medicine Supply Chain', In 2020 21st Asia-Pacific Network Operations and Management Symposium (APNOMS), pp. 235-238, IEEE, September 2020.

[79] A. Musamih, K. Salah, R. Jayaraman, J. Arshad, M. Debe, Y. Al-Hammadi, S. Ellahham, 'A Blockchain-Based Approach for Drug Traceability in Healthcare Supply Chain', IEEE Access, 9, 9728-9743, 2021

[80] I. A. Omar, R. Jayaraman, M. S. Debe, K. Salah, I. Yaqoob, M. Omar, 'Automating procurement contracts in the healthcare supply chain using blockchain smart contracts', IEEE Access, 9, 37397-37409, 2021.

[81] S. K. Panda, S. C. Satapathy, 'Drug traceability and transparency in medical supply chain using blockchain for easing the process and creating trust between stakeholders and consumers', Personal and Ubiquitous Computing, 1-17, 2021.

[82] Y. Kong, Y. Wang, '"Blockchain Plus Supply Chain Finance" Boosts Pharmaceutical Industry: A Case Study of Jixiangtian Blockchain Medical and Health Service Platform', In E3S Web of Conferences, Vol. 292, EDP Sciences, 2021.

[83] G. Saldamli, V. Reddy, K. S. Bojja, M. K. Gururaja, Y. Doddaveerappa, L. Tawalbeh, 'Health Care Insurance Fraud Detection Using Blockchain', In 2020 Seventh International Conference on Software Defined Systems (SDS), pp. 145-152, IEEE, April 2020.

[84] W. Liu, Q. Yu, Z. Li, Z. Li, Y. Su, J. Zhou, 'A blockchain-based system for anti-fraud of healthcare insurance', In 2019 IEEE 5th International Conference on Computer and Communications (ICCC), pp. 1264-1268, IEEE, December, 2019.

[85] B. Alhasan, M. Qatawneh, W. Almobaideen, 'Blockchain Technology for Preventing Counterfeit in Health Insurance', In 2021 International Conference on Information Technology (ICIT), pp. 935-941, IEEE, July 2021.

[86] M. H. Chinaei, H. H. Gharakheili, & V. Sivaraman, 'Optimal Witnessing of Healthcare IoT Data Using Blockchain Logging Contract', IEEE Internet of Things Journal, 2021.

[87] M. Thenmozhi, R. Dhanalakshmi, S. Geetha, R. Valli, 'Implementing blockchain technologies for health insurance claim processing in hospitals', Materials Today: Proceedings, 2021.

[88] A. Goyal, A. Elhence, V. Chamola, B. Sikdar, 'A Blockchain and Machine Learning based Framework for Efficient Health Insurance Management', In Proceedings of the 19th ACM Conference on Embedded Networked Sensor Systems, pp. 511-515, November 2021.

[89] Y. Huang, J. Wu, & C. Long, 'Drug ledger: A practical blockchain system for drug traceability and regulation', In 2018 IEEE International Conference on Internet of Things (iThings) and IEEE Green Computing and Communications (GreenCom) and IEEE Cyber, Physical and Social Computing (CPSCom) and IEEE Smart Data (SmartData), pp. 1137-1144, IEEE, July 2018.

[90] W. Xie, B. Wang, Z. Ye, W. Wu, J. You, Q. Zhou, 'Simulation-based blockchain design to secure biopharmaceutical supply chain', In 2019 Winter Simulation Conference (WSC), pp. 797-808, IEEE, December 2019.

[91] X. Liu, A. V. Barenji, Z. Li, B. Montreuil, G. Q. Huang, 'Blockchain-based smart tracking and tracing platform for drug supply chain', Computers & Industrial Engineering, 161, 107669, 2021.

[92] M. Uddin, 'Blockchain Medledger: Hyperledger fabric enabled drug traceability system for counterfeit drugs in pharmaceutical industry', International Journal of Pharmaceutics, 597, 120235, 2021.

[93] J. H. Tseng, Y. C. Liao, B. Chong, S. W. Liao, 'Governance on the drug supply chain via gcoin blockchain', International journal of environmental research and public health, 15(6), 1055, 2018.

[94] P. Sylim, F. Liu, A. Marcelo, P. Fontelo,, 'Blockchain technology for detecting falsified and substandard drugs in distribution: pharmaceutical supply chain intervention', JMIR research protocols, 7(9), e10163, 2018.

[95] K. Abbas, M. Afaq, T. Ahmed Khan, W. C. Song, 'A blockchain and machine learning-based drug supply chain management and recommendation system for the smart pharmaceutical industry', Electronics, 9(5), 852, 2020.

[96] E. Ehioghae, S. Idowu, O. Ebiesuwa, 'Enhanced Drug Anti-Counterfeiting and Verification System for the Pharmaceutical Drug Supply Chain using Blockchain', International Journal of Computer Applications, 975, 8887, February 2021.

[97] W. Chien, J. de Jesus, B. Taylor, V. Dods, L. Alekseyev, D. Shoda, P. B. Shieh, 'The last mile: DSCSA solution through blockchain technology: drug tracking, tracing, and verification at the last mile of the pharmaceutical supply chain with BRUINchain', Blockchain in Healthcare Today, 2020.

[98] S. Shamshad, K. Mahmood, S. Kumari, C. M. Chen, 'A secure blockchain-based e-health records storage and sharing scheme', Journal of Information Security and Applications, 55, 102590, 2020.

10

Fusion of Machine Learning and Blockchain Techniques in IoT-based Smart Healthcare Systems

Navita[1] and Pooja Mittal[2]

[1]Ph.D Scholar, Department of Computer Science and Applications,
Maharshi Dayanand University Rohtak, India;
Email: navitamehra55@gmail.com
[2]Assistant Professor, Department of Computer Science and Applications,
Maharshi Dayanand University Rohtak, India;
Email: mpoojmdu@gmail.com

Abstract

Smart healthcare is one of the rising areas of research in collaboration with smart techniques like sensors, the Internet of Things (IoT), and various information analytics techniques to convey effective healthcare services to the patients at a lesser amount of cost. The healthcare system faces lots of problems in dealing with a large volume of data generated day by day, and processing and exploring useful information from it. Such a large volume of data is generated through various IoT-based medical gadgets, electronic health records, wearable sensors, telemedicine, and mobile health services that may demand effective, fast, and secure data analysis techniques to offer high-quality services to patients. To provide such types of services, machine learning and blockchain techniques are highly appreciated in various domains. This study mainly discusses the categorization of powerful machine learning and blockchain techniques highly effective to process such a vast amount of data. As the healthcare system encompasses highly sensitive data regarding patient health as well as personal information which must require appropriate security services, blockchain technology can be used as an effective solution for offering security services to highly sensitive data. Machine learning and blockchain technology together provide lots of opportunities for the healthcare system to attain all its goals like reduced healthcare cost, effective diagnosis, timely treatment, offering transparency in regulatory reporting, and effective health data management. The main aim of this chapter is to provide a detailed description of the relationship between blockchain and machine learning techniques to empower an IoT-based healthcare system. Along with that, this chapter also discusses the major issues and challenges faced while implementing IoT-based smart healthcare systems by collaborating machine learning and blockchain techniques.

10.1 Introduction

Healthcare is a crucial part of healthy life. Unfortunately, the gradually increasing elderly population and continuous increase in chronic diseases have placed significant stress on the recent healthcare system [1], and the request for healthcare services like beds, nurses, and doctors is increasing at a fast rate [2]. The healthcare system requires an urgent solution to decrease the strain on the healthcare system while ensuring to offer superior quality service to the patient at a high level. Most of the researchers have worked toward it and found an alternative solution to deal with such problems [3]. Recently, the healthcare system has gradually benefited from various innovative techniques such as the Internet of Things (IoT), blockchain, cloud computing, and machine learning (ML) techniques to support the smart healthcare system to improve the disease diagnosis and treatment process. Internet of Things (IoT) has been gradually recognized as an efficient resolution to reduce the pressure on the healthcare system and is considered the most recent area of research. ML techniques in the healthcare system offer smart health monitoring services and medical automation services in different aspects and the environment (like home, office, hospitals, etc.). Integration of cloud and IoT in the healthcare system offers connectivity between real-time applications. IoT devices have the capability of collecting and communicating a large quantity of health data generated day by day by different medical sensors either worn or implanted on a patient's body [4]. The extraction and analysis of meaningful information from the large quantity of data collected by different medical sensors and devices required human intervention. But the introduction of ML- and AI-based techniques offers intelligent automation through which prediction and analysis would be possible without human intervention [5]. The traditional prognosis and diagnosis framework was weakly designed which could hamper the appropriate identification and treatment of patients. The incorporation of IoT and ML has provided an intelligent and smart decision support healthcare system that supports in identifying disease at an initial phase and further improve the lifetime of the patient. The ML strategies have made a critical involvement in the analysis of diseases in medical services and therefore offer a vital decrease in specialist appointment costs and provide an overall improvement in the quality of patient care. The incorporation of IoT and ML in the healthcare system offers lots of possibilities in improving the healthcare system by reducing the stress points of the recent healthcare system. But when health data need to be stored and exchanged, maintaining the privacy of data is a major concern, and the current healthcare data storage system lacks top-tier security [6]. In such circumstances, blockchain could be the possible solution to susceptibilities like data theft and hacking. Blockchain technology integrated with the feature of interoperability that offers secure exchange of medical data among all the entities and systems involved that results in lots of benefits like saving time, improving communication, and improving operational efficiency [7], [8]. According to a report, it is projected that the universal blockchain technology in the healthcare market value would be more than $1.636 billion by 2025 [9].

This chapter explains the fusion of ML and blockchain technology in smart IoT-based healthcare systems. Section 10.2 describes the work done by different researchers in this domain, Section 10.3 discusses major issues and challenges faced while establishing IoT in healthcare, Section 10.4 represents the idea of blockchain technology in healthcare, Section 10.5 describes the introduction of AI and ML in the healthcare system, Section 10.6 explains the working process of the smart healthcare system, Section 10.7 explains all possible ML algorithms utilized in the healthcare system, and Section 10.8 will describe all possible solutions offered by blockchain and ML techniques to a smart healthcare system. Finally, Section 10.9 concludes the chapter.

10.2 Literature Review

Due to the highly threatening risk of a pandemic (COVID-19), the World Health Organization (WHO) has recommended each nation to create a "pandemic plan". A pandemic plan is frequently planned as per the WHO's pandemic stages, determined to accomplish unambiguous outcomes in pandemic administration from the beginning [10]. Different types of techniques have been recognized for any emergency in healthcare. Every disaster is categorized into four stages: mitigation, arrangement, response, and recovery [11]. The "table top exercise or discussion-based session" is an important strategy used to find an effective solution to deal with any crisis; such discussion will create a scenario that is assisted by both communication and collaboration among different areas and sectors such as management, laborers, money, etc. A legitimate study could provide a broad framework and a psychological model that reproduces the perfect decision-making environment [11]. Among all important technologies, IoT is considered as one of the most common technologies to deal with such type of pandemic which can provide connectivity among all the devices on a network [12], [13]. Exchange of information performed between each object of the network is authorized through a cycle dependent on the agreement conveyed across all hubs (that is, the gadgets/clients). The idea of blockchain originates from the way that each person holds a reference to the past one utilizing a cryptographic strategy. Blockchain is not put away on a concentrated server like ordinary web-based administrations, rather on network gadgets (PCs) called hubs, every one of which has a duplicate of the whole blockchain [14]. The replication and capacity of various duplicates of various blockchains across network hubs guarantee more noteworthy framework security and value among clients, who can get to similar data simultaneously, and consequently the detectability and changeability of the approved exchanges contained in the squares. Therefore, blockchain is a distributed organization where all organization clients can trust the framework without confiding in each other. Along with blockchain technology, machine learning (ML) is also utilized for the analysis of a large amount of data collected through different sensors and medical devices. Different researchers have worked on ML-based healthcare monitoring systems to offer effective and timely treatment to patients by offering smart real-time health monitoring systems. Ani *et al.* [15] presented an IoT- and ML-based healthcare monitoring system for the prediction of chronic diseases like asthma, cancer, diabetes, etc., for elderly people. The proposed system will consist of three layers: (1) hardware layer (microcontroller, blood pressure, and other sensor-based devices); (2) application layer (cloud server and web environment); (3) different ML algorithms like naïve Bayes, RFT, etc. Almost all RFTs provide an accuracy of 93%. Onasanya *et al.* [16] presented a cancer-based system using the concept of IoT and cloud services. This model is based on the concept of a body wireless sensor network (BWSN) connected to a patent to monitor their health parameters. Kumar *et al.* [17]

suggested a healthcare monitoring framework for the identification of heart disease using IoT and ML. The proposed system is composed of three layers. Layer 1 gathered the data from sensors implanted on the patient's body, layer 2 utilized HBase to collect a large volume of information, and layer 3 focuses on data analytics techniques known as ML. Varadarajan *et al.* [18] proposed an IoT-based healthcare system using the cloud as a secure storage space. They proposed a fuzzy-rule-based technique for the diagnosis of disease. The proposed system is composed of eight components named UCI repository dataset, medical sensors, cloud computing, fuzzy temporal neural classifier, and data accumulation. The proposed system has been implemented in JAVA programming using the Amazon cloud. K-NN, DT, SVM, and NB were utilized as important ML techniques, and each obtained a different accuracy result. Verma *et al.* [19] discussed a smart system for the diagnosis of student diseases. The proposed methodology works in three stages. At stage 1, data is collected from different sensors, in stage 2, data is moved to the cloud by using the specific gateway, and in stage 3, processing of data is done. Along with that, a warning system will also be attached to warn the patients and caregivers in case of any emergency. Gutte *et al.* [20] presented an IoT-based health monitoring system using different IoT devices. The proposed system is divided into four tiers. Tier-1 works on the data gathering concept, Tier-2 processes the data using Raspberry Pi as a microcontroller, and Tier-3 relates to the data storage concept. The proposed system sends an SMS and email to the patient or caregivers in case any critical situation arises.

From the study of work done by different researchers, we come to know that both blockchain and ML techniques play an efficient role in offering secure and smart healthcare monitoring services to patients in real time at a low cost.

10.3 Issues and Challenges While Establishing IoT In Healthcare

Although IoT has a lot of benefits and is capable to resolve a wide variety of complications in different areas, still there exist lots of challenges. These complications might be in the term of defeating safety issues, protection concerns, and so on [21], [22], [23], [24], [25]. This section briefly describes all possible issues and challenges: major challenges faced while establishing IoT are depicted in Table 10.1.

Table 10.1   Major problems and challenges of IoT

Issues	Challenges	Explanations
Safety issue	Design practices Rate vs. safety	Deficiency of assets in training people about safety-based IoT plan in the near future Deficiency of information about the cost and benefit occurs during IoT analysis
	Standard and metrics Confidentiality, authentication and control Field upgradeability Device uselessness	Deficiency of knowledge about ethics and standard measures related to security issues Lack of optimality-controlled role may cause hijacking and cyber-attack problem Challenges arise due to a lack of knowledge about the maintainability and upgradeability of IoT devices Limited inferences on replacement of the old and undesirable gadgets
Interoperability issues	Proprietary ecosystem and consumer wish lack Technical risk Configuration	Lack of knowledge about data collection format may cause interoperability issues; so proper security protocols must be implemented Less awareness about the technical issue may generate technical issues Issues arise due to a lack of knowledge about standard configurations used for multiple gadgets
IoT ethics issue	The explosion of ideal efforts	The challenge arises due to the lack of efforts made in generating standard protocols
Emerging economy issue	Investment	The limited investment made in the field of researching IoT and its related activities
Developmental issues	Infrastructure resources Technical and industrial developments	Causes lots of burden on communication and Internet infrastructure globally Fewer efforts are done in empowering the Internet and communication infrastructure

10.4 Involvement of Blockchain Technique in the Healthcare System

Most of the developed countries have been using the concept of blockchain technology in healthcare since 2012 to enable the secure transfer of patient health records and support a secure supply chain and help healthcare researchers in this domain. The blockchain security feature improves health data security. This technique offers a public ID or key to each patient and a private key used to open the record only when required. This technique would limit the hijacking of patient personal or sensitive information [7]. The blockchain healthcare management system is shown in Figure 10.1 and blockchain healthcare facilities are represented in Figure 10.2.

10.4.1 Securing and tracking health supplies

Blockchain can help to ensure a secure and distinguishable way to track all medical supplies in a completely transparent manner. Along with that, it also keeps track of labor costs and emissions of carbon linked with the construction of these items.

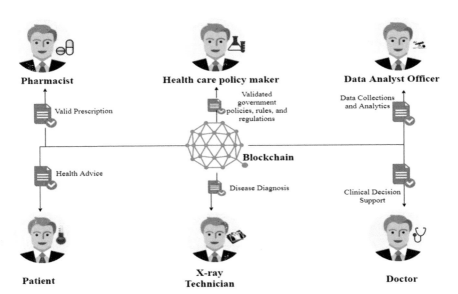

Figure 10.1 Blockchain-based healthcare management system [26].

10.4.2 Storing health information

There has been considerable advancement in the storage of patient health reports without affecting their integrity and consistency; still, a loss of health data is a normal situation, and with the increasing number of hackers all around the world, hacking of this information has turned into the most common procedure of penetrating and overshadowing past strategies. Meditab [28] is the best software product company that claims to be one of the foremost software development companies for medical records. For supporting data seeking facility, electronic faxes are utilized by the company to transmit medical records. But, sometimes, this way of data transfer might be unsafe or unreliable.

10.4.3 Remote patient monitoring

Remote patient monitoring is an innovation to empower checking of patients outside of hospitals, for example, in home, office, or in a far region, which may improve access to healthcare services and may decrease healthcare costs. RPM involves the constant remote care of patients by healthcare service providers. Many devices like smartwatches, smart band, etc., enabled smart technologies to monitor various health parameters. Such types of monitoring systems are useful in rural sectors in order to provide healthcare services in highly isolated areas.

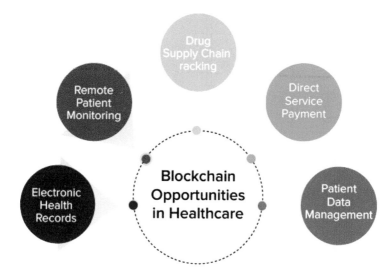

Figure 10.2 Blockchain opportunities in healthcare [27].

10.4.4 Disease outbreak and tracking

One of the unique characteristics of blockchain technology is to support real-time disease reporting and identification of disease patterns which could help in finding the origin of disease and its various transmission factors.

10.5 Indulgement of AI and ML in Healthcare System

ML, a subpart of AI, has been extensively utilized in the healthcare domain. Data regarding patient health, hospitals treatment reports, and health status need to be digitalized and a large amount of data has originated in the field of healthcare due to the advancement of information and communication technology (ICT) in healthcare [29], [30]. Machine learning is effectively utilized for medical purposes to analyze highly complex healthcare data becoming the most important topic of research for everyone. This study will describe how the healthcare business sector employs different machine learning techniques including neural network learning, deep learning, and feature fusion techniques to accomplish data analysis tasks as shown in Figure 10.3.

The main objective of this study is to appreciate health monitoring, human activity detection, disease diagnosis, and prediction. Khafajiy *et al.* [31] demonstrated a real-time elderly people health monitoring system. Many elderly people who live alone suffer from various physiological disorders. The developed system is highly effective in sensing any physiological disorders between people and taking immediate action regarding patient health. The proposed system is highly effective in collecting and manipulating data. Biag *et al.* [32] experimented on 120 ECG monitoring systems and found out that ECG is considered as one of the most effective for monitoring vital parameters regarding patient health and is generally used to monitor and measure the high-risk factor related to patient health. Banaee *et al.* [33] provided a review on the application of different machine learning algorithms utilized for analyzing health-related data highly effective in analyzing the health-related data generated from different wearable sensors that may be worn or implanted on the patient body. Gumaei *et al.* [34] utilized different MLs to identify human activity. They developed a multisensory-based hybrid model using the deep learning concept. The proposed approach intelligently processes the data collected from different sensors for the care of elderly and ill people in a smart manner.

Ali *et al.* [36] presented a disease prediction system by utilizing the concept of information gain and feature fusion technique to forecast the

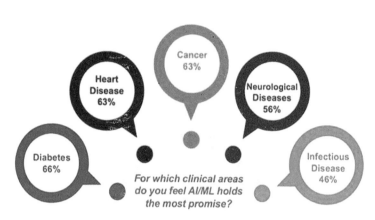

Figure 10.3   Most common chronic health disease benefited by ML/AI [35].

existence of heart disease. Juutinen *et al.* [37] utilized different machine learning algorithms for the forecast of Parkinson's disease using inertial sensors inserted in smartphones. Inertial sensors are highly effective for measuring walking parameters. This system classifies the disorders of movement based on signals captured by sensors. They worked on three feature selection and five classification algorithms to identify Parkinson's disorder among patients.

10.5.1 Supervised learning

Supervised learning comes under the category of ML techniques whose training data contains the labeled value. The training data may contain multiple inputs with one labeled output. The labels are utilized in the machine learning models to assist themselves during training to make an accurate prediction for new data. Classification and regression come under the category of supervised learning. For medical purposes, classification is a common issue. In most clinical cases, a clinician diagnoses the illness of any patient based on symptoms. The regression method is typically utilized for forecasting numerical outcomes, for example, predicted number of days for which the patient has to stay in hospitals depending upon the collected dataset regarding vital signs.

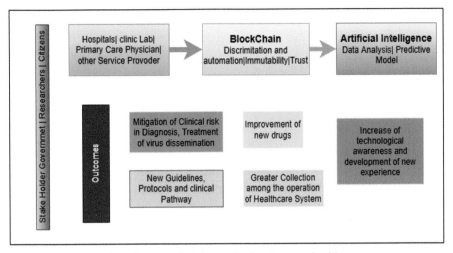

Figure 10.4 AI and blockchain architecture in smart healthcare system.

Decision tree (DT), random forest tree (RFT), linear and logistic regression (LR), and support vector machine (SVM) all come under supervised learning techniques [38]. Random forest tree also comes under the category of ensemble classifier that can be made up of multiple decision trees that have been trained individually. The RFT build-up from multiple decision trees well operates with noisy data [39].

10.5.2 Unsupervised learning

Unsupervised ML works with unlabeled data to find the pattern within an existing dataset [39]. This algorithm typically works on the method of grouping data on the basis of similarity. *K*-means clustering is the most well-known and widely utilized method. The association task related to clustering is also solved with the help of such a clustering method. The standard *k*-means clustering was further modified and named *k*-medians or *k*-medoids. In the case of *k*-clusters, the nearest center and mean are calculated based on Euclidean distance. The most common problem associated with this method is that it needs a probable number of means [40].

10.5.3 Semi-supervised learning

In such types of ML techniques, the model is trained by utilizing labeled and unlabeled datasets. In such datasets, the amount of unlabeled data is very large as compared to the labeled dataset. In such a learning technique, the first unsupervised learning approach is followed to cluster the unlabeled dataset, and, after that, they are labeled by utilizing the existing labeled dataset. Mostly in healthcare, such type of learning approach is followed for speech analysis and protein classification sequences [41].

10.5.4 Reinforcement learning

In such types of ML approach, a model learns from the moments of activities within an environment rather than by openly trained with the given set of inputs and outputs. Such types of trained machines learn from previous experiences and take an immediate new decision based on trial and error learning. This technique is highly utilized to generate expert systems for healthcare applications [41].

10.6 Working Process

ML process mainly describes the way followed by any ML algorithm for the implementation of any data. ML process starts with the collection of data from multiple sources and is followed by a tweaked interaction of information; after that, the information would be fed to any ML algorithm based on the statement of the problem. The complete working process is divided into five stages as discussed below. Before the application of any ML algorithm, the objective for which the ML is to be applied must be clear. For finding the solution to any problem, first of all, data must be collected and the below strategies must be applied appropriately [42], [43].

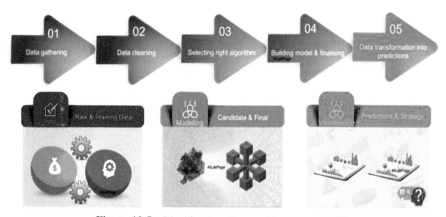

Figure 10.5 Machine learning working process [43].

- Data collection: This phase collects the data from multiple sources, be it the sensor collected or recorded data depending upon the nature of the application. In terms of smart healthcare, data may be collected from different worn and implanted sensors on the patient's body or some other medical device.
- Data cleaning: Data collected from different sources may consist of noise and some irregularities. The data cleaning process starts with the removal of noise and inconsistency among data to find a better result.
- Selecting right algorithm: As there exist lots of ML algorithms and each algorithm is evaluated with different evaluation measures like root mean square error, confusion matrix, sensitivity, error rate, and time-consumption. Depending upon these evaluation parameters, the best algorithm is selected.
- Building model: Building a model means training the model using the training dataset. Training dataset is computed from original dataset by splitting it into training and testing dataset using some splitting criteria.
- Prediction: Prediction mainly refers to the outcomes generated after the application of the model.

10.7 Machine Learning Algorithms in the Healthcare Industry

ML algorithms are highly employed in the healthcare industry for the forecast of diverse diseases and are described below.

10.7.1 Support vector machine

This is the most common ML algorithm highly utilized in the healthcare industry. This algorithm can classify the hidden information by deriving the nominated features and by creating a high-dimensional hyperplane to classify the data into different classes to derive a decision model. Due to the ability to classify the high-dimensional data by utilizing a minimum training set of features, it is highly utilized for handling physiological data for healthcare applications. This technique is enabled by kernel techniques; various optimization problems in high dimensional space utilized kernel functions. During SVM implementation, training data is plotted onto a new feature space by using the kernel optimization function and produces a large marginal separation between the training data in the novel feature space [44], [45]. For example, a given labeled sequence $(A_1, B_1), \ldots (A_m, B_m)$, where A represents the covariant and $B \; \mathcal{E} \; \{-1,1\}$, is the answer. A kernel function k is operated through SVM by means of using the following equation:

$$f(x) = a^T x + c. \tag{10.1}$$

At this point, a denotes the dimensional coefficient and c is the offset.

SVM has fewer overfitting complications and can simply work with composite structured datasets.

In the healthcare sector, normal wellbeing boundaries considered by SVM are HR, ECG, and SpO_2 which are short and annotated health parameters. Hu *et al.* [46] utilized the SVM technique to diagnose arrhythmia disease from ECG signals captured from ECG devices. Similarly, Lee *et al.* [47] also proposed an SVM method for identifying arrhythmia using ECG signals. The result displayed that the SVM method with polynomial kernel provides an improved outcome as compared to other kernels. Most of the researchers have used SVM for the classification and diagnosis of different diseases on both sensor and clinically recorded data. However, from the conclusion made by different researchers, they found that SVM cannot be used as an appropriate technique to integrate domain knowledge and symbolic knowledge seamlessly on data collected from sensors. Along with that, SVM cannot be applied to find unforeseen info from unlabeled data.

10.7.2 Logistic regression (LR)

LR is a highly flexible and robust method used for dichotomous classification purposes. This method is highly effective to make a classification in terms of binary outcomes or states like yes or no, success or failure, and will occur or not occur. This method is highly effective in most cases where linear regression does not work. This is because along with making numerical value predictions, it also predicts a dichotomous value. Due to this associated feature, this method is named logistic regression. In most healthcare problems, this method provides the best result; for example, it will answer a question like whether or not a person suffers from a specific disease or not based on the symptom [48].

Logistics regression is exceptionally valuable when a dependent variable is categorical. Logit function utilized for assessment of likelihood is given by the following equation:

$$p^\wedge = h_\emptyset(x) = \sigma(x^T\theta) \tag{10.2}$$

where

$$\sigma(t) = \frac{1}{1+e^{-t}}.$$

10.7.3 Decision tree (DT)

DT is highly effective to discriminate between the selected features; this technique is considered as one of the most significant learning techniques due to effective representation of rule classification. This method follows a tree-like structure whose internal nodes are represented by the attributes and branch denotes the decision rules, and finally, outcomes are represented by the leaf nodes. This method is highly effective to support both binaries as well as multi-class classification problems. For splitting the nodes, entropy is used as a splitting criterion represented by eqn (10.3) [33], [40], [49]. The issue related to the decision tree is that it cannot deal with direct connections among information [50]

$$\text{Entropy}(S) = \sum_{i=1}^{n} -p_i * \log2(p_i) \tag{10.3}$$

where P_i is the likelihood of S belonging to class i and n is the quantity of classes.

DT is one of the most reliable methods to be used for medical data classification. This method effectively works with multivariate sensors due to independent levels in the decision tree. One of the common problems associated with a decision tree is that such type of classification approach may suffer from an overfitting problem.

10.7.4 Random forest tree (RFT)

RFT can be considered as the advanced version of the decision tree. This method is highly effective for multiple predictions due to its high accuracy and its capability to handle a large number of features with small samples. This method works on a synthesis system to syndicate the severity level of chronic diseases mainly for ECG and HR (heart rate). As this method generates a forest of decision trees, selection of the best tree is done based on threshold rules. In such a method, one of the most robust features has been for the initial splitting of data in a tree-like structure. Random forest is utilized for making sickness risk predictions and to analyze ECG and MRI data based on the health history of patients. This method is highly effective as compared to the decision tree because it overcomes the overfitting problem [33].

10.7.5 Discriminant analysis (DA)

DA is an ML method utilized for estimating the accuracy of the classification of one object into one or two categories. This method is highly utilized for medical business purposes mainly for early detection of diabetic peripheral to refine the analytical features of blood vessels imagining. This method is also utilized to detect the major symptom of psychological strength disorientation and for electronic health record organization systems [51].

10.7.6 K-nearest neighbor

K-NN is the simplest classification method that is completely non-parametric. For example, we have provided a point A_0 that we want to classify into one out of k-groups; so first of all, this method observed the k data points that are close to the A_0. One of the most common advantages associated with this method is its simplicity. There are only two major choices: number of neighbors denoted as "k" and the distance metric to be followed by either Euclidian distance or Manhattan distance. The number of neighbors is selected by the test validation method or by applying the classifier on a new test dataset [39].

10.8 Solution via Blockchain and Artificial Intelligence

In most developed countries, healthcare can be counted as one of the significant portions of gross domestic product (GDP). On the other hand, expenditure in the healthcare sector continues to increase due to inefficient procedures and health data breach. In such a situation, blockchain technology has the capability to upgrade things. This technique can offer a wide range of services like offering security services by encrypting healthcare data and epidemic management. Estonia is a pioneer in this field, having executed blockchain innovation in medical care in 2012. At present, blockchain is utilized to keep up with its total medical care charging framework, 95% of wellbeing information, and 99% of solution data [40].

In what manner blockchain technology can become a problematic solver?

1. Each client that has been verified will have a duplicate of the common record. This will address the information assortment issue. The AI models can be taken care of straightforwardly with amazingly reliable information, and the outcomes can be removed.
2. Genuine information can be utilized to prepare the model(s). Thus, the productivity and precision of models will improve, bringing the extra expense down to the focal power.
3. The patient can get assistance on the most proficient method to live a healthy life. The model can be prepared to utilize advice given to different patients with comparative issues or manifestations (by clinicians).
4. At any point, when a patient poses a basic inquiry about their wellbeing, a prepared model with a test is presented on the blockchain network; the model can anticipate flare-ups and make recommendations to specialists. Normal language processing can be utilized to analyze disease and make treatment suggestions.
5. The prepared model can make clinical recommendations to clinicians dependent on the symptoms of the patient.
6. The model has been trained to predict outbreaks. For instance, if a patient has a test and the outcomes are transferred to the blockchain network, the model can foresee an epidemic and make proposals to the specialist.
7. To accomplish any clinical trial, we utilize a variation of machines and apparatus in the medical service industry.

Each machine, or part of a machine, has a particular life expectancy. It can likewise expect when the machine or a piece of the machine should be transformed or taken out.

Conclusion

The growing utilization of blockchain and AI in smart healthcare systems has attracted the attention of most researchers and developers. IoT researchers and developers are collaborating to merge distinctive emerging technology named IoT, AI, and blockchain to highly benefit society in possible ways like providing smart infrastructure, healthcare, smart city, smart agriculture, etc. However, further improvements in the existing system are possible only if the possible problems and shortcomings in the existing system are considered at present. This article will detail existing issues and challenges in our healthcare system and also provides the best approach in which the two most emerging technologies blockchain and ML can be combined to enhance the efficiency of the existing healthcare system. By including the emerging technologies, the new system can provide a secure, transparent, and intelligent smart healthcare system named the Internet of Medical Things (IoMT).

Acknowledgement

I would like to express my gratitude and appreciation to all those who gave me the possibility to complete this chapter. Special thanks to my supervisor Dr. Pooja Mittal whose help, stimulating suggestions, and encouragement helped me in all times of the fabrication process and in writing this chapter.

References

[1] Australian Institute of Health and Welfare. (2014). Australia's Health. [Online]. Available: http://www.aihw.gov.au/WorkArea/ Download Asset. aspx-?id=60129548150

[2] E. Perrier, 'Positive Disruption: Healthcare, Ageing and Participation in the Age of Technology, Sydney, NSW, Australia: The McKell Institute, 2015.

[3] S. B. Baker et al.,' IoT for Smart Healthcare: Technologies, Challenges, and Opportunities ', vol. 5, 2017, 2169-3536, IEEE Access.

[4] L. Greco, G. et al.,' Trends in IoT based solutions for healthcare: Moving AI to the edge', Pattern Recognition Letter, 135, pp. 346–353, 2020.

[5] Agrawal, V. et al., 'Hyperglycemia prediction using machine learning: A probabilistic approach, International Conference on Advances in Computing and Data Sciences, Springer, Singapore, pp. 304–312, 2019.

[6] M.M. Hassan, et al., 'Future of the Internet of Things Emerging with Blockchain and Smart Contracts', International Journal of Advanced Computer Science and Applications, 11, 2020.

[7] R.B. Fekih et al., 'Application of Blockchain Technology in Healthcare: A Comprehensive Study', The Impact of Digital Technologies on Public Health in Developed and Developing Countries 12157, 268 – 276, 2020.

[8] N. M. Kumar and P. K. Mallick, 'Blockchain technology for security issues and challenges in IoT', Procedia Computer Science, vol. 132, pp. 1815–1823, 2018.

[9] Bills, Claims and Analysis: Blockchain in Healthcare Business Intelligence, 2019.

[10] D. T. Jamison et al., 'Disease Control Priorities, 3rd edition: improving health and reducing poverty, Disease Control Priorities, Vol. 9, 2017.

[11] B. Zarzaur, 'Blueprint for restructuring a department of surgery in concert with the health care system during a pandemic: The University of Wisconsin Experience', JAMA surgery 2020.

[12] D. Niewolny, 'How the Internet of Things is revolutionizing Health care', a white paper by Healthcare Segment Manager, Free scales Semiconductor.

[13] D. Miorandi et al.,' Internet of Things: vision, applications and research challenges', ELSEVIER, pp. 1497-1516I, April 2012.

[14] R.B. Fekih,' Application of Blockchain Technology in Healthcare: A Comprehensive Study', The Impact of Digital Technologies on Public Health in Developed and Developing Countries, 12157, 268 – 276, 2020.

[15] R. Ani et al.,' IoT based patient Monitoring and diagnostic prediction tool using ensemble classifier', International Conference on Advances in Computing, Communications and Informatics (ICACCI), IEEE, pp. 1588-1593,2017.

[16] Onasanya et al., 'IoT Implementation for Cancer care and Business Analytics/Cloud Services in Healthcare System', In proceeding of 10th International Conference on Utility and Cloud Computing, pp. 205-206, 2017.

[17] Kumar et al., 'A Novel Three-Tier Internet of Things Architecture with machine learning algorithms for early detection of heart diseases', Computers and Electrical Engineering, vol.65, 222-235, 2018.

[18] Varatharajana et al., 'Cloud and IoT based disease prediction and diagnosis system for healthcare using Fuzzy neural classifier', Future Generation Computer System, vol. 86, pp. 527-534, 2018.

[19] Verma et al.,' Cloud-Centric IoT based disease diagnosis healthcare framework', Journal of parallel and distributed computing, vol.116, pp. 27-38, 2018.

[20] Gutte et al., 'IoT based health monitoring system using Rasberry Pi', Fourth International Conference on Computing Control and Automation (ICCUBEA), pp. 1-5, 2018.

[21] R. Pranav et al., 'Application of Blockchain and Internet of Things in Healthcare and Medical Sector: Applications, Challenges, and Future Perspectives, Hindawi Journal of Food Quality, 2021.

[22] Nallapaneni Manoj Kumar et al.,' Blockchain technology for security issues and challenges in IoT', International Conference on Computational Intelligence and Data Science (ICCIDS 2018) Procedia Computer Science 132, 1815–1823, 2018.

[23] A. Dorri, 'Blockchain for IoT security and privacy: the case study of a smart home', in Proceedings of the 2017 IEEE International Conference on Pervasive Computing and Communications Workshops (PerCom Workshops), IEEE, Kona, USA, March 2017.

[24] R. Porkodi and V. Bhuvaneswari, 'Internet of Things (IoT) applications and communication enabling technology standards: an overview', in Proceedings of the 2014 International Conference on Intelligent Computing Applications, IEEE, Washington, DC, USA, March 2014.

[25] B. Le Nguyen et al., 'Privacy-preserving blockchain technique to achieve secure and reliable sharing of IoT data', Computers, Materials and Continua, vol. 65, no. 1, pp. 87–107, 2020.

[26] blockchain management system overview - Google Search

[27] Blockchain in Healthcare: The Challenges, Applications, and Benefits (mobileappdaily.com)

[28] Best EMR Software Company | Medical Software | EHR | IMS | Meditab. https://www.meditab.com/. (Accessed on 10/28/2021).

[29] N. V. Pardakhe et al.,' Machine Learning and Blockchain Techniques Used in Healthcare System', 2019 IEEE Pune Section International Conference (Pune Con) MIT World Peace University, Pune, India, 2019.

[30] G. Kumar et al., 'A survey on machine learning techniques in health care industry', International Journal of Recent Research Aspects 3, 128–132, 2016.

[31] M. Al-Khafajiy, et al., 'Remote health monitoring of elderly through wearable sensors. Multimed Tools Appl 78, 24681–24706, 2019.

[32] M. Biag et al., 'A comprehensive survey of wearable and wireless ECG monitoring systems for older adults, Med. Biol. Eng. Comput. 51(5):485–495, 2013.

[33] H. Banaee et al., 'Data mining for wearable sensors in health monitoring systems: A review of recent trends and challenges', Sensors, 13(12):17472–17500, 2013.

[34] A. Gumaei et al., 'Hybrid Deep Learning Model for Human Activity Recognition Using Multimodal Body Sensing Data', IEEE Access 2019, 7, 99152–99160, 2019.

[35] AI In Healthcare infrastructure, 2020, AI In Healthcare infrastructure - BLOCKGENI

[36] F. Ali et al., 'A Smart Healthcare Monitoring System for Heart Disease Prediction Based On Ensemble Deep Learning and Feature Fusion', Information Fusion, 63, 2020.

[37] M. Juutinen et al., 'Parkinson's disease detection from 20-step walking tests using inertial sensors of a smartphone: Machine learning approach based on an observational case-control study ', PLoS ONE 15(7),2020.

[38] scikit-learn: machine learning in Python — scikit-learn 1.0.1 documentation. https://scikit-learn.org/stable/. (Accessed on 10/28/2021).

[39] sklearn.Ensemble.Random Forest Classifier — scikit-learn 1.0.1 documentation. https://scikit-learn.org/stable/modules/ generated/sklearn. ensemble.RandomForestClassifier.html. (Accessed on 10/28/2021).

[40] Blockchain Applications in Healthcare. https://www.news-medical. net/health/Blockchain-Applications-in-Healthcare.aspx. (Accessed on 10/28/2021).

[41] N. Aggarwal et al.,' Machine Learning Applications for IoT Healthcare', Machine Learning Approaches for Convergence of IoT and Blockchain, (129–144), Scrivener Publishing LLC, 2021.

[42] M. Hassan et al., 'A Blockchain-Based Intelligent Machine Learning System for Smart Health Care' Preprints, 1, 2021.

[43] Vinod Sharma's Blog How Machine Learning Algorithms Works: An Overview - Vinod Sharma's Blog (google.com)

[44] S. Liu et al., 'Computational and Statistical Methods for Analysing Big Data with Applications, Academic Press, Pages 57-85, 2016.

[45] Xin-SheYang, 'Support Vector Machines and Regression', Introduction to algorithms for Data Mining and Machine Learning, 129-138, 2019.

[46] F. Hu et al.,' Medical ad hoc sensor networks (MASN) with wavelet-based ECG data mining', *Ad Hoc Robust Netw.*, **6**:986–1012, 2008.

[47] K.H. Lee et al, 'Low-energy formulations of support vector machine kernel functions for biomedical sensor applications', J. *Signal Process. Syst.*, **69**:339–349, 2012.

[48] E.B Seufert, 'Quantitive Methods for Product Management ', in Freemium Economics, 2014.

[49] S. Aich, et al. ,' A nonlinear decision tree-based classification approach to predict the Parkinson's disease using different feature sets of voice data', International Conference of Advanced Communication and Technology, 638–642, 2018.

[50] Asif-Ar-Raihan et al.,' Performance Evaluation and comparative analysis of different machine learning algorithms in predicting cardiovascular disease', Journal of Engineering Latter, 29 (2).

[51] S. Yang, et al.,' Early Detection of Disease Using Electronic Health Records and Fisher's Wishart Discriminant Analysis, Procedia Computer Science, Volume 140, Pages 393-402, 2018.

Index

About the Editors

Vishal Jain is presently working as an Associate Professor with Sharda University, Greater Noida, UP, India. Before that, he has worked for several years as an Associate Professor with Bharati Vidyapeeth's Institute of Computer Applications and Management (BVICAM), New Delhi, India. He has more than 14 years of experience in academics. He obtained the Ph.D. (CSE), M.Tech (CSE), MBA (HR), MCA, MCP, and CCNA. He has more than 370 research citation indices with Google Scholar (h-index score 9 and i-10 index 9). He has authored more than 70 research papers in reputed conferences and journals, including Web of Science and Scopus. He has authored and edited more than 10 books with various reputed publishers, including Springer, Apple Academic Press, CRC, Taylor and Francis Group, Scrivener, Wiley, Emerald, and IGI-Global. His research areas include information retrieval, semantic web, ontology engineering, data mining, *ad hoc* networks, and sensor networks. Dr. **Jain** received a Young Active Member Award for the year 2012–2013 from the Computer Society of India, Best Faculty Award for the year 2017, and Best Researcher Award for the year 2019 from BVICAM, New Delhi.

Jyotir Moy Chatterjee is an Assistant Professor with the Information Technology Department, Lord Buddha Education Foundation, Kathmandu, Nepal & Young Ambassador 2020-2021 Egyptian Scientific Research Group (SRGE). His research interests include machine learning and deep learning. He is serving as an Editorial Board Member of various reputed journals of IGI Global and serving as a Reviewer for various reputed journals & international conferences of IGI Global, Elsevier, Springer, IEEE, Tech Science Press, etc.

Ishaani Priyadarshini is a faculty with the School of Information, University of California, Berkeley, USA. She received the Ph.D. and master's degrees from the University of Delaware. She has authored several book chapters for reputed publishers and is also an author to several publications for SCIE indexed journals. As a certified reviewer, she conducts peer review of research papers for prestigious IEEE, Elsevier, and Springer journals and is a part of the Editorial Board for *International Journal of Information Security and*

Privacy (IJISP). Dr. Priyadarshini's areas of research include cybersecurity, applied machine learning, and artificial intelligence.

Fadi Al-Turjman received the Ph.D. degree in computer science from Queen's University, Canada, in 2011. He is the Associate Dean for research and the Founding Director of the International Research Center for AI and IoT at Near East University, Nicosia, Cyprus. Prof. Al-Turjman is the head of the Artificial Intelligence Engineering Department, and a leading authority in the areas of smart/intelligent IoT systems, wireless, and mobile networks' architectures, protocols, deployments, and performance evaluation in Artificial Intelligence of Things (AIoT). His publication history spans over 350 SCI/E publications, in addition to numerous keynotes and plenary talks at flagship venues. He has authored and edited more than 40 books about cognition, security, and wireless sensor networks' deployments in smart IoT environments, which have been published by well-reputed publishers such as Taylor and Francis, Elsevier, IET, and Springer. Dr. Al-Turjman has received several recognitions and best papers' awards at top international conferences. He also received the prestigious Best Research Paper Award from Elsevier Computer Communications Journal for the period 2015–2018, in addition to the Top Researcher Award for 2018 at Antalya Bilim University, Turkey. Prof. Al-Turjman has led several international symposia and workshops in flagship communication society conferences. Currently, he serves as Book Series Editor and the lead Guest/Associate Editor for several top tier journals, including the *IEEE Communications Surveys and Tutorials* (IF 23.9) and the Elsevier Sustainable Cities and Society (IF 5.7), in addition to organizing international conferences and symposiums on the most up to date research topics in AI and IoT.